FREQUENCY AND TRANSIENT CHARACTERISTICS OF MMIC TRANSMISSION LINES, CIRCUITS AND ANTENNAS

by

Constantine A. Balanis
and
Anastasis C. Polycarpou
Seong-Ook Park
James P.K. Gilb
Michael R. Lyons

Arizona State University

with additional revisions by
Wexford Press

Wexford
2008

TABLE OF CONTENTS

iv

LIST OF TABLES

LIST OF FIGURES

vii

CHAPTER 1

INTRODUCTION

1.1 Objectives

The objectives of this book are to develop analytical formulations, computational techniques, and computer algorithms to analyze the frequency and transient characteristics of Monolithic Microwave Integrated Circuits (MMIC's) and lines. In particular, the following:

    a. Two-dimensional planar microwave integrated transmission lines and circuits with isotropic and anisotropic substrates.

    b. Three-dimensional electronic packaging circuits.

    c. Aperture antennas loaded with dielectric and ferrite materials.

To accomplish these objectives, advanced computational electromagnetic methods were used. In particular, the methods dvanced and used were:

    a. Method of Moments (MoM) in the spetral domain, referred to as the Spectral Domain Approach (SDA).

    b. Vector-based Finite Element Method (FEM).

    c. Hybrid technique based on the Finite Element Method and Method of Moments (FEM/Mom).

The main topics addressed here, each one in a separate chapter, are the following:

    a. Two-dimensional planar transmission lines and circuits.

    b. Three-dimensional electronic packaging circuits.

c. Analytical asymptotic extraction technique for multi-layer transmission lines and planar circuits.

d. Dielectric- and ferrite-loaded, cavity-backed aperture and microstrip antennas.

1.2 Summary

The most important topics are arranged by chapter heading. A more extensive summary of each of these topics can be found at the start of each of these chapters.

1.2.1 Two-dimensional planar transmission lines and circuits

Full-wave numerical techniques, such as the Finite Element Method (FEM) and the Spectral Domain Approach (SDA), were developed and optimized for the analysis of 2-D microwave and millimeter wave structures. Propagation characteristics including effective dielectric constant, attenuation constant, and characteristic impedance are calculated versus frequency for the dominant and higher-order modes. Anisotropic materials such as sapphires, PTFE cloth and ferrites are also modeled and analyzed using the FEM. Substrate compensation is used in multi-layer, multi-conductor circuits to minimize coupling, pulse distortion, and equalize the phase velocities of the even and odd modes for edge-coupled microstrip directional couplers with tighter couplings and better performance. In addition, a closed-form expression for the effective dielectric constant of an open microstrip line on a two-layer substrate is derived which can be used to reduced the computation time in CAD procedures.

1.2.2 Three-dimensional electronic packaging circuits

A three-dimensional vector finite element code was developed to simulate practical electronic packaging circuits including microstrip/coplanar discontinuities, spiral inductors, filters, conducting vias, *etc.* The packaging structures may be shielded or open, depending on the application. Open structures are modeled using efficient absorbing boundary conditions to terminate the finite element mesh. The circuit is usually excited with the dominant mode of an infinite transmission line, although other modes of propagation may be chosen as an excitation. The specific modal distribution at the input port is determined using a full-wave 2-D eigenvalue analysis.

The formulation and respective code were extensively verified by comparing magnitude and phase of S-parameters with data obtained using other numerical techniques, such as the Finite-Difference Time-Domain (FDTD) method. Comparisons between the two methods clearly demonstrate the effectiveness, accuracy and versatility of the finite element method in predicting propagation characteristics of high-frequency packages.

1.2.3 Analytical asymptotic extraction techniques for multi-layer transmission lines and planar circuits

The main contributions of this work are the development of analytical solutions for the evaluation of the asymptotic matrix elements of planar transmission lines and circuits using the Method of Moments. The objective of this effort was to reduce the computation time to evaluate the impedance matrix elements. Analytical techniques have been successfully developed and used to improve the computational efficiency, in some cases by a factor of 50, while retaining or even improving the accuracy of the results. This has been demonstrated successfully in a number of problems; specifically, planar transmission lines and printed circuits including microstrip dipoles, symmetric and asymmetric gap discontinuities, and scattering from a rectangular microstrip patch.

1.2.4 Dielectric- and ferrite-loaded, cavity-backed aperture and microstrip antennas

A hybrid Finite Element Method/Method of Moment (FEM/MoM) approach was used in the analysis and design of cavity-backed aperture antennas loaded with dielectric and/or ferrite materials. The antenna is treated both as a scatterer and a radiator. In the first case, the analysis is concentrated on predicting the radar cross section of the antenna versus frequency or angle. In the second case, the antenna is excited using a standard coaxial line. A novel coaxial feed model is implemented using the finite element method. Comparisons of input impedance data with measurements illustrate the inherent accuracy and efficiency of the proposed feed model.

In addition to multiple dielectric layers, the cavity volume of the antenna may be loaded with magnetized ferrites. Variation of the externally biased field results in a significant shift in resonant frequency, thereby tuning the antenna over a wide bandwidth. This tuning effect is illustrated with numerical predictions for the case of a square aperture backed by a ferrite-loaded rectangular cavity.

1.3   Journal Publications

1. Gilb, J. P. K. and Balanis, C. A.,"MIS slow-wave structures over a wide range of parameters," *IEEE Trans. Microwave Theory Tech.*, vol. 40, no. 12, pp. 1248-2154, Dec. 1992.

2. Gilb, J. P. K. and Balanis, C. A.,"Accurate and efficient computation of dielectric losses in multi-level, multi-conductor microstrip for CAD applications," *IEEE Trans. Microwave Theory Tech.*, vol. 41, no. 3, pp. 527-530, Mar. 1993.

3. Lyons, M. R., Gilb, J. P. K. and Balanis, C. A.,"Enhanced dominant mode operation of a shielded multilayer coplanar waveguide via substrate compensation," *IEEE Trans. Microwave Theory Tech.*, vol. 41, no. 9, pp. 1564-1567, Sept. 1993.

4. Park, S.-O and Balanis, C. A.,"Efficient Kernel calculation of cylindrical antennas," *IEEE Trans. Antennas Propag.*, vol. 43, no. 11, pp. 1328-1331, Nov. 1995.

5. Lyons, M. R. and Balanis, C. A.,"Transient coupling reduction and design consideration in edge-coupled coplanar waveguide couplers," *IEEE Trans. Microwave Theory Tech.*, vol. 44, no. 5, pp. 778-783, May. 1996.

6. Polycarpou, A. C, Lyons M. R., and Balanis, C. A., "A two-dimensional finite element formulation of the perfectly matched layer," *IEEE Microwave and Guided Wave Letters*, vol. 6, no. 9, pp. 338-340, Sep. 1996.

7. Polycarpou, A. C, Lyons M. R., and Balanis, C. A., "A full-wave finite element analysis of coplanar waveguides with anisotropic substrates," *IEEE Trans. Microwave Theory Tech.*, vol. 44, no. 10, pp. 1650-1663, Oct. 1996.

8. Park, S.-O and Balanis, C. A.,"Dispersion of open microstrip lines using closed-form asymptotic extraction," *IEEE Trans. Microwave Theory Tech.*, vol. 45, no. 3, Mar. 1997.

9. Park, S.-O and Balanis, C. A.,"Dispersion of coupled microstrip using closed-form asymptotic extraction," *IEEE Microwave and Guided Wave Letters*, vol. 7, no. 3, pp.84-86, Mar. 1997.

10. Park, S.-O and Balanis, C. A.,"Analytical technique to evaluate the asymptotic part of the impedance matrix of Sommerfeld-type integrals," *IEEE Trans. Antennas Propag.*, vol. 45, no. 5, May 1997.

11. Polycarpou, A. C, Tirkas, P. A. and Balanis, C. A.,"The finite element method for modeling three-dimensional electronic packaging circuits," submitted for publication in the *IEEE Trans. Microwave Theory Tech.*.

12. Park, S.-O, Balanis, C. A. and Birtcher, C. R.,"Analytical evaluation of the asymptotic impedance matrix of a grounded dielectric slab with roof-top function," submitted for publication in the *IEEE Trans. Antennas Propag.*.

13. Polycarpou, A. C., Lyons M. R. and Balanis, C. A., "On the accuracy of perfectly matched layers using a finite element formulation," submitted for publication in the *IEEE Trans. Microwave Theory Tech.*.

1.4   Conference Publications

1. Gilb, J. P. K. and Balanis, C. A.,"Coupling reduction in high-speed, high-density digital interconnects with substrate compensation," *Topical meeting on electrical performance of Electronic Packaging*, pp. 116-118, Apr. 22-24, 1992, Tucson, AZ.

2. Gilb, J. P. K. and Balanis, C. A.,"Fast and accurate computation of dielectric losses in multi-layer, multi-conductor microstrip structures,"*1992 IEEE International Microwave Symposium Digest*, pp. 385-388, Jun. 1-5, 1992.

3. Gilb, J. P. K. and Balanis, C. A.,"MIS slow-wave structures over a wide range of parameters," *1992 IEEE International Microwave Symposium Digest*, pp. 877-880, Jun. 1-5, 1992.

4. Gilb, J. P. K. and Balanis, C. A.,"Wide-band reduction of coupling and pulse distortion in multi-conductor integrated circuit lines," Invited paper, *1992 Ultra-Wideband Short-Pulse Electromagnetic Conference*, Oct. 8-10, 1992, Polytechnic University, Brooklyn, NY.

5. Gilb, J. P. K. and Balanis, C. A. and Lyons, M. R.,"Design tools for substrate compensated low-coupling structures," *1992 Microwave Hybrid Circuits Conference*, Oct. 25-28, 1992, Wickenburg, AZ.

6. Balanis, C. A. and Gilb, J. P. K.,"Coupling and distortion in multi-conductor, high-speed digital interconnects," Invited paper, Workshop on Pico-second and Fempto-second Electromagnetic Pulses-Analysis and Applications, *1993 IEEE International Microwave Symposium Digest*, Jun. 14-18, 1993, Atlanta, GA.

7. Gilb, J. P. K. and Balanis, C. A.,"Closed-form expressions for the design of coupled microstrip lines with two substrate layers," *1993 IEEE International Microwave Symposium Digest*, pp. 1005-1008, Jun. 14-18, 1993, Atlanta, GA.

8. Lyons, M. R., Gilb, J. P. K. and Balanis, C. A.,"Enhanced dominant mode operation of shielded multilayer coplanar waveguide," *1993 IEEE International Microwave Symposium Digest*, pp. 943-946, Jun. 14-18, 1993, Atlanta, GA.

9. Gilb, J. P. K. and Balanis, C. A.,"Improved performance of microstrip couplers through multi-layer substrates," *23rd European Microwave Conference*, Sept. 6-9, 1993, Madrid, Spain.

10. Lyons, M. R. and Balanis, C. A.,"Transient coupling reduction in edge-coupled coplanar waveguide forward directional couplers," *1994 IEEE MTT-S International Microwave Symposium Digest*, vol. 3, pp. 1685-1688, May 23-27, 1994, San Diego, CA.

11. Tirkas, P. A., El-Ghazaly, S., Rajan, S., El-Sharawy, E. and Balanis, C. A.,"Characterization of high performance packages: electrical, thermal, and stress issues," *1995 National Radio Science Meeting - U.S. National Committee of URSI*, Boulder, CO, Jan. 1995.

12. Lyons, M. R., Polycarpou, A. C. and Balanis, C. A., "On the Accuracy of Perfectly Matched Layers Using a Finite Element Formulation," *1996 IEEE MTT-S International Microwave Symposium*, pp. 205-208, Jun. 17-21, 1996, San Francisco, CA.

13. Polycarpou, A. C., Lyons, M. R. and Balanis, C. A., "Dispersive Effects of a Thin Metal-Insulating Layer in MMIC Structures," *1996 IEEE MTT-S International Microwave Symposium*, pp. 303-306, Jun. 17-21, 1996, San Francisco, CA.

14. El-Ghazaly, S. M., Megahed, M. A., Balanis, C. A., and Blakey, P. A., "Progress of electromagnetic modeling of device and circuit interactions", *Progress in Electromagnetics Research Symposium (PIERS) '96*, p. 73, Jul. 8-12, 1996, Innsbruck, Austria.

15. Toupikov, M., Pan, G.-W, and Balanis, C. A., "On the application of weighted wavelet expansion to surface integral equations", *Progress in Electromagnetics Research Symposium (PIERS) '96*, p. 583, Jul. 8-12, 1996, Innsbruck, Austria.

16. Polycarpou, A. C., Lyons, M. R., Aberle J. and Balanis, C. A., "Analysis of arbitrary shaped cavity-backed patch antennas using a hybridization of the finite element and spectral domain methods, "*IEEE Antennas and Propagation Society International Symposium*, pp. 130-133, Jul. 21-26, 1996, Baltimore, MD.

17. Park, S.-O and Balanis, C. A., "Closed-form extraction of dispersion characteristics of open microstrip lines", *1996 URSI Radio Science Meeting*, Jul. 21-26, 1996, Baltimore, MD.

18. Polycarpou, A. C. and Balanis, C. A., "Analysis of ferrite loaded cavity backed slot antennas using a hybrid FEM/MoM approach", "*IEEE Antennas and Propagation Society International Symposium*, Jul. 13-18, 1997, Montreal, Canada.

19. Polycarpou, A. C. and Balanis, C. A., "Analysis of microwave integrated circuits using the finite element method", "*IEEE Antennas and Propagation Society International Symposium*, Jul. 13-18, 1997, Montreal, Canada.

1.5 Theses and Dissertations

1. Lyons, M. R., "Pulse Distortion and Coupling Control Using Substrate Compensation in Multi-Layer Coplanar Waveguides", MS(EE) Thesis, Arizona State University, August 1994.

2. Park, S.-O., "Analytical Techniques for the Evaluation of Asymptotic Matrix Elements in Electromagnetic Problems", Ph.D.(EE) Dissertation, Arizona State University, May 1997.

3. Polycarpou, A. C., "The Finite Element Method for the Analysis of Microwave Integrated Circuits and Aperture Antennas", Ph.D.(EE) Dissertation, Arizona State University, August 1997.

4. Gilb, J. P. K., "Transient and Frequency Domain Analysis of Discontinuities and Multi-Conductor Interconnects in Multi-Level Microstrip", Ph.D.(EE) Dissertation, Arizona State University, December 1997.

CHAPTER 2

ANALYSIS AND DESIGN OF 2-D PLANAR TRANSMISSION LINES AND CIRCUITS

2.1 Finite Element Analysis

This chapter presents a full-wave 2-D finite element analysis of planar microwave transmission lines and circuits with possible anisotropic substrates. Dispersive parameters such as propagation constant and characteristic impedance are calculated at every frequency for each governing modal distribution. This formulation results in a generalized eigenvalue matrix system with the unknown eigenvector representing the transverse and longitudinal electric fields, whereas the unknown eigenvalue representing the propagation constant. The computer code written based on the present formulation is interfaced with another code in a following chapter for the analysis of 3-D packaging structures. The 2-D code is implemented to obtain the frequency dependent modal distribution at the input port, which is later used as the excitation field for the 3-D microwave circuit. The characteristic impedance and propagation constant for all the transmission lines at the ports are also computed using the 2-D finite element code.

2.1.1 Introduction

Accurate prediction of the propagation characteristics in planar structures using isotropic and anisotropic substrate materials is essential in the design of Monolithic Microwave Integrated Circuits (MMIC's) [1]-[5]. Since many substrate materials of interest in microwave and millimeter wave applications exhibit dielectric and/or magnetic anisotropy (such as sapphires, ceramics and ferrites), the effects due to variations in the material parameters must be fully accounted for. Principal axis rotations of anisotropic substrates in MMIC's might also lead to significant variations in effective dielectric constant and characteristic impedance, especially at microwave and millimeter wave frequencies. The dispersive characteristics of coplanar waveguides (CPWs) and other planar structures on single and multi-layer isotropic substrates have been extensively analyzed in the literature [6]-[10]. To an extent, the effects due to anisotropy have been examined using methods other than FEM; however primarily only for uniaxial and/or biaxial substrates [1],[3],[5]. Axis rotations in various planes, which introduces off-diagonal elements in the permittivity and/or permeability tensors, were previously investigated using the SDA [2],[4].

Although quasi-static methods have been employed to analyze the dominant mode characteristics of CPWs and other planar structures on isotropic and anisotropic substrates, this yields accurate results only at very low frequencies [11]. More accurate frequency dependent solutions have been obtained using full-wave analyses such as the Spectral Domain Approach (SDA) [2],[3],[10] and the Finite Element Method (FEM) [12]. While the SDA is a popular choice for analyzing regular planar structures, the FEM is the most generally applicable and versatile, since it is possible to model arbitrary geometric and material complexities. When using the FEM, the domain of interest is discretized using simple geometric shapes, such as triangles, where the fields are approximated using linear or higher order basis functions. Because of this, it is also relatively straightforward to compute quantities of interest in MMIC transmission lines, such as total power, voltage difference and characteristic impedance. A major drawback of the FEM is the appearance of non-physical or spurious modes. However, these non-physical solutions to Maxwell's equations, which appear when using nodal-based finite elements, can be avoided using vector-based finite elements [12]. In addition to imposing tangential continuity of the electric and magnetic fields across element boundaries, the vector finite elements also satisfy the divergence-free condition. Using this type of element, the obtained numerical solutions correspond to the true physical solutions of the structure. Allocation of computer resources is also a major concern when using the FEM since such a technique requires storage and manipulation of large sparse systems. In this case, sparse linear solvers are usually more suitable than direct solvers [13].

Until recently, most finite element formulations have been used to analyze the propagation characteristics of isotropic and biaxially anisotropic waveguides [13],[14] with explicit application to only isotropic microstrip structures [12]. In this chapter, an extended vector-based finite element formulation for *biaxial and transverse plane anisotropic materials* is presented and used to characterize shielded 2-D microwave structures. A numerically efficient algorithm for finding the largest eigenvalue and eigenvector is presented based on a forward iteration approach. Higher eigenpairs can be found using a Gram-Schmidt orthogonalization process [15]. In addition, an explicit formulation for calculating characteristic impedance applicable to slot-like MMIC structures is given for linear triangular finite elements. Numerical results are compared with existing published data to verify the finite element code.

2.1.2   Eigen problem formulation

A full-wave analysis of shielded microwave circuits, which accommodates both electric and magnetic anisotropic materials, is fully described by the electric field Helmholtz's equation given by

$$\nabla \times \left(\bar{\bar{\mu}}_r^{-1} \cdot \nabla \times \mathbf{E} \right) - k_o^2 \, \bar{\bar{\epsilon}}_r \, \mathbf{E} = 0 \qquad (2.1)$$

The permittivity and permeability tensors are modeled as 3×3 matrices of the following form:

$$\bar{\bar{\epsilon}}_r = \begin{bmatrix} \epsilon_{xx} & \epsilon_{xy} & 0 \\ \epsilon_{yx} & \epsilon_{yy} & 0 \\ 0 & 0 & \epsilon_{zz} \end{bmatrix} \qquad\qquad \bar{\bar{\mu}}_r = \begin{bmatrix} \mu_{xx} & \mu_{xy} & 0 \\ \mu_{yx} & \mu_{yy} & 0 \\ 0 & 0 & \mu_{zz} \end{bmatrix}.$$

The inverse permeability tensor is defined as

$$\bar{\bar{\mu}}_r^{-1} = \begin{bmatrix} \mu_{xx}^{inv} & \mu_{xy}^{inv} & 0 \\ \mu_{yx}^{inv} & \mu_{yy}^{inv} & 0 \\ 0 & 0 & \mu_{zz}^{inv} \end{bmatrix}.$$

The latter is evaluated in closed form using symbolic manipulations of the original permeability tensor. The representative variational functional for the Helmholtz's equation in a two-dimensional domain Ω can be expressed as

$$F(\mathbf{E}) = \frac{1}{2} \int \int_{\Omega} \left[(\nabla \times \mathbf{E}) \, \bar{\bar{\mu}}_r^{-1} \, (\nabla \times \mathbf{E})^* - k_o^2 \mathbf{E} \, \bar{\bar{\epsilon}}_r \mathbf{E}^* \right] d\Omega. \tag{2.2}$$

Assuming that the dependence of the fields in the $z-$direction is $e^{-jk_z z}$, the functional in (2.2) can be discretized, using a similar procedure as in [12], to obtain the following generalized eigenvalue matrix system

$$\begin{bmatrix} A_{tt} & 0 \\ 0 & 0 \end{bmatrix} \begin{Bmatrix} e_t \\ e_z \end{Bmatrix} = -k_z^2 \begin{bmatrix} B_{tt} & B_{tz} \\ B_{zt} & B_{zz} \end{bmatrix} \begin{Bmatrix} e_t \\ e_z \end{Bmatrix} \tag{2.3}$$

Examining (2.3), it is observed that the matrix on the left is singular; therefore, there are N eigenvectors that correspond to an eigenvalue k_z of zero. N is equal to the number of nodes in the finite element mesh. These solutions are non-physical since they do not satisfy the Helmholtz's equation. However, they can be avoided by re-writing the matrix system in the following form:

$$\begin{bmatrix} B_{tt} & B_{tz} \\ B_{zt} & B_{zz} \end{bmatrix} \begin{Bmatrix} e_t \\ e_z \end{Bmatrix} = \frac{k_{max}^2}{k_{max}^2 - k_z^2} \begin{bmatrix} B_{tt} + \frac{A_{tt}}{k_{max}^2} & B_{tz} \\ B_{zt} & B_{zz} \end{bmatrix} \begin{Bmatrix} e_t \\ e_z \end{Bmatrix} \tag{2.4}$$

where k_{max} is the maximum propagation constant in the longitudinal direction. This is defined as $k_{max}^2 = k_o^2 \epsilon_{max} \mu_{max}$; ϵ_{max} and μ_{max} are respectively the maximum permittivity and permeability of the domain. In microwave circuit applications, people are usually interested in the few most dominant modes of the structure; $i.e.$, the ones that correspond to the most positive k_z^2 and therefore, a larger value of $k_{max}^2/(k_{max}^2 - k_z^2)$. Thus, for positive values of k_z, the eigenvalue $k_{max}^2/(k_{max}^2 - k_z^2)$ ranges from 1 to "infinity"; 1 corresponds to $k_z = 0$ and "infinity" corresponds to $k_z = k_{max}$. As a result of this transformation, the zero eigenvalues are shifted outside the range of interest.

It is also important to realize here that although the generalized eigenvalue matrix system was derived based on a propagating wave of the form $e^{-jk_z z}$, there is no restriction on whether k_z is real or complex. One could easily assume that $k_z = \beta - j\alpha$ where α is really the attenuation constant and β is the actual propagation constant in the longitudinal direction. Thus, lossy as well as anisotropic material can be analyzed using the approach described in this section.

The obtained generalized eigenvalue matrix system can be solved using either a standard direct solver or an iterative solver. The former usually results in the computation of all the eigenvalues and eigenvectors of the matrix system. However, in practice, only the first few dominant modes are desired; therefore, an iterative solver is usually more appropriate. In this particular case, a power forward iteration is used.

A. Characteristic impedance formulation

After solving the eigenvalue problem given in (2.3) at each frequency point, the propagation constant in the $z-$direction and the corresponding normalized transverse and longitudinal fields inside the structure can be obtained. Both the propagation constant and the fields are needed for the calculation of the characteristic impedance. Although the definition of the characteristic impedance is not unique for inhomogeneous waveguide structures, the voltage-power definition was chosen for the current analysis. The corresponding expression is given by [5]

$$Z_c = \frac{V V^*}{2 P} \quad , \qquad P = \sum_{i=1}^{N} P_i \qquad (2.5)$$

where V is the voltage difference between the transmission line and the ground plane, P_i is the power calculated in each element, P is the total power flowing in the $z-$direction, and N is the total number of finite elements in the domain of interest. The elemental power P_i is computed using the Poynting vector defined as

$$P_i = \frac{1}{2}\mathbf{Re}\left[\int\int_{\triangle} \mathbf{E}_i \times \mathbf{H}_i^* \cdot \hat{a}_z ds\right] = \frac{1}{2}\mathbf{Re}\left[\int\int_{\triangle}\left(E_x H_y^* - E_y H_x^*\right) ds\right] \qquad (2.6)$$

where the magnetic field components H_x and H_y are calculated directly from Maxwell's equations; note that the expressions for the electric field components are known in closed form. The above integration is performed over the area of each finite element.

B.   A generalized eigenvalue solver

The solution of a generalized eigenvalue problem defined as

$$[K]\{x\} = \lambda[M]\{x\} \qquad (2.7)$$

can be computationally intensive and time demanding, especially as the number of unknowns increases. There are various methods of solving for both the eigenvalues and the corresponding eigenvectors. The simplest method is to store both matrices in a full format and then use a direct solver, like those available in EISPACK. Such a solver usually computes all the eigenvalues and eigenvectors of the matrix system. This approach, however, is very inefficient both in terms of computational time and memory requirements. One of the most suitable methods for solving a generalized eigenvalue problem is the power iteration, otherwise known as forward and inverse power iteration. Note that the forward power iteration is used to estimate the largest eigenvalues of the matrix system, whereas the inverse power iteration is used to estimate the lowest eigenvalues. The major advantages to using a power iteration are: first, speed-up in computational time; second, complete utilization of the sparsity of the matrices; third, computation of only a selected number of eigenvalue/eigenvector pairs. As far as the latter is concerned, it is important to realize that in analyzing waveguide structures, only the most dominant modes are significant; therefore, it is not necessary that higher-order eigenvalues and eigenvectors are calculated. In addition, the accuracy of the higher-order eigenvalues and eigenvectors deteriorates accordingly, which is another reason for not calculating more than a few eigenmodes. It is also important to say that the forward power iteration converges much faster than the inverse power iteration [12], at least for the type of problems considered in this study, which is another significant reason for implementing this particular algorithm.

2.1.3 Validation and numerical results

A complete FEM code based on the analytical formulation presented in previous sections was written and tested for a variety of geometries and materials. The FEM code was interfaced with SDRC I-DEAS, a software package from Structural Dynamics Research Corporation with preprocessing requirements such as meshing, material definition, and boundary conditions. It was also interfaced with other well-known packages such as PLOTMTV, GEOMVIEW, TECPLOT, and GNUPLOT which can be used for data visualization and important geometry checks.

The first geometry considered in validating the finite element formulation and corresponding code was the coupled microstrip line, shown in Fig. 2.1, initially examined by Mostafa *et al.* [4] using a spectral domain approach. The coupled microstrip line rests on a uniaxial boron nitride substrate ($\epsilon_{xx} = \epsilon_{zz} = 5.12$, $\epsilon_{yy} = 3.4$). The effective dielectric constant, ϵ_{reff}, versus a principal axis crystal rotation angle, θ, as defined in [4], is depicted in Fig. 2.2 for two different frequencies: $f = 10$ GHz and $f = 20$ GHz. A comparison between our results and data obtained from [4] shows a very good agreement between the two methods.

A second geometry considered was a unilateral finline, illustrated in Fig. 2.3, already examined by Mansour *et al.* [5]. The frequency dependence of the effective dielectric constant ϵ_{reff} and the characteristic impedance Z_c is illustrated in Fig. 2.4. Data obtained from the corresponding figure in [5] are also shown on the same graph. The agreement between the two sets of data is very good. It is important to mention here that the accuracy of the characteristic impedance depends on the mesh density in the finite element region, especially near the slot. The reason is due to the fact that the formulation of the characteristic impedance involves the actual field distribution in the entire structure. Specifically, fields in the vicinity of the slot, which are known to exhibit rapid spatial variations, require either a finer mesh or higher order elements for good representation. In obtaining the results shown in Fig. 2.4, 13 linear triangular elements were used across the slot and a total of 1036 elements in

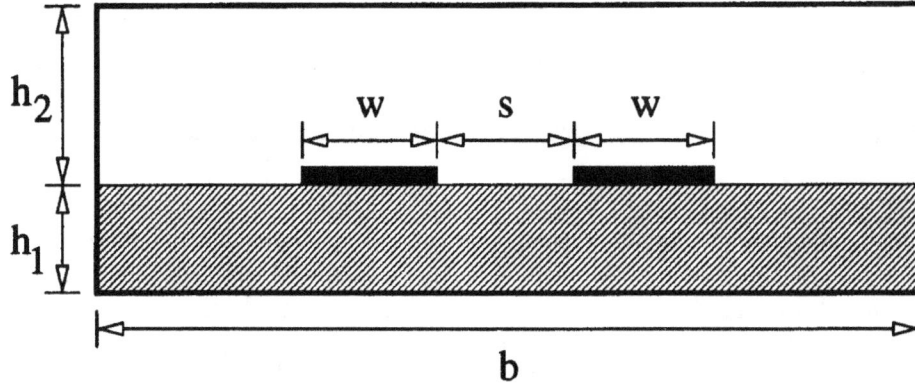

Fig. 2.1: Geometry of a coupled microstrip line. Dimensions: $h_1 = 1.5$ mm, $w = s = 1.5$ mm, $b = 8.5$ mm.

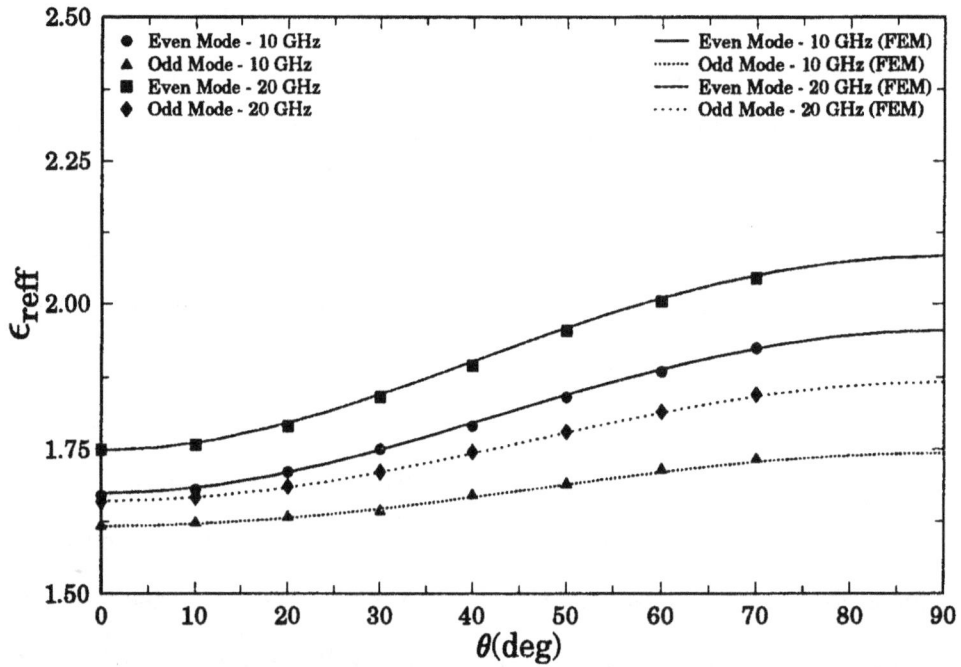

Fig. 2.2: Dispersion curves for the dominant mode of a coupled microstrip line on a boron nitride substrate ($\epsilon_{xx} = 5.12$, $\epsilon_{yy} = 3.4$, $\epsilon_{zz} = 5.12$). The markers represent data extracted from a paper by Mostafa *et al.*.

the entire structure.

A suspended coplanar waveguide with magnetic anisotropic uniaxial substrates, shown in Fig. 2.5, was also analyzed using the finite element code. This geometry was initially examined by Mazé-Merceur *et al.* [3] using the spectral domain approach. Defining the quantity $\epsilon_{reff} \cdot \mu_{reff} = (k_z/k_o)^2$ [3], the effects due to both electric and magnetic anisotropy are accounted for. A comparison of the $\epsilon_{reff} \cdot \mu_{reff}$ versus frequency between the two methods is illustrated in Fig. 2.6 for a variety of isotropic and uniaxially magnetic anisotropic substrates. A good agreement is observed between the two sets of data.

2.2 Spectral Domain Approach

The Spectral Domain Approach (SDA) is a full-wave technique that is using a modified Galerkin's approach applied in the Fourier transform or spectral domain. The SDA is by far the most popular choice of solving boundary-value problems involving simple and uniform planar printed circuits such as CPW, microstrip, and fin-line, to name a few. Computation time is a major concern in the calculation of complex electromagnetic problems, and one major advantage to using the SDA is the excellent numerical efficiency it provides. This is mainly due to the flexibility in choice of basis functions used. Proper selection can enable use of fewer basis function expansion terms for numerical convergence yielding faster computer codes [16]. However, a drawback of the SDA is its limited use, particularly to regularly shaped structures such as planar surfaces.

In order to decrease the numerical expense involved with other full-wave approaches without sacrificing accuracy, the SDA has been applied to non-uniform microstrip lines [17]-[31]. As with uniform microstrip lines, the SDA is perhaps the simplest and most efficient full-wave approach, combining a straightforward derivation with efficient expansions for the current densities. In addition, because the SDA is a full-wave method, the effects of surface waves, radiation, and coupling between

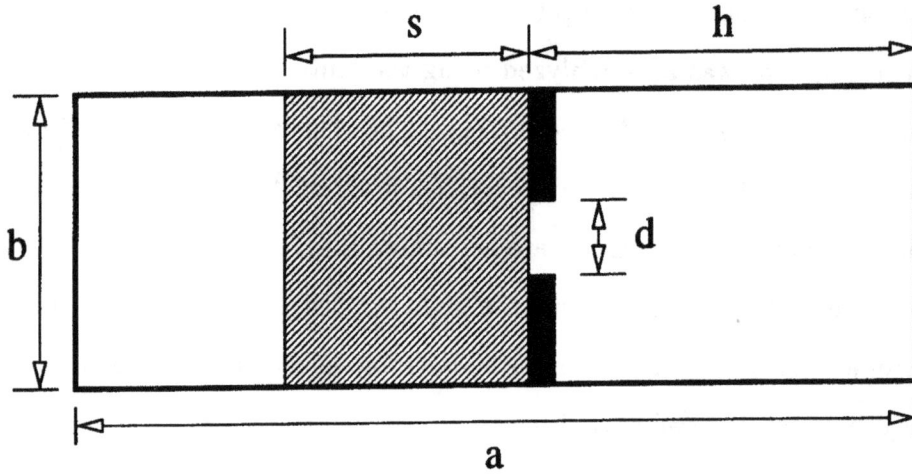

Fig. 2.3: Geometry of a unilateral finline on a dielectric substrate of $\epsilon_r = 3.8$. Dimensions: $a = 2b = 4.7752$ mm, $s = 0.127$ mm, $h = 2.3876$ mm, $d = 0.47752$ mm.

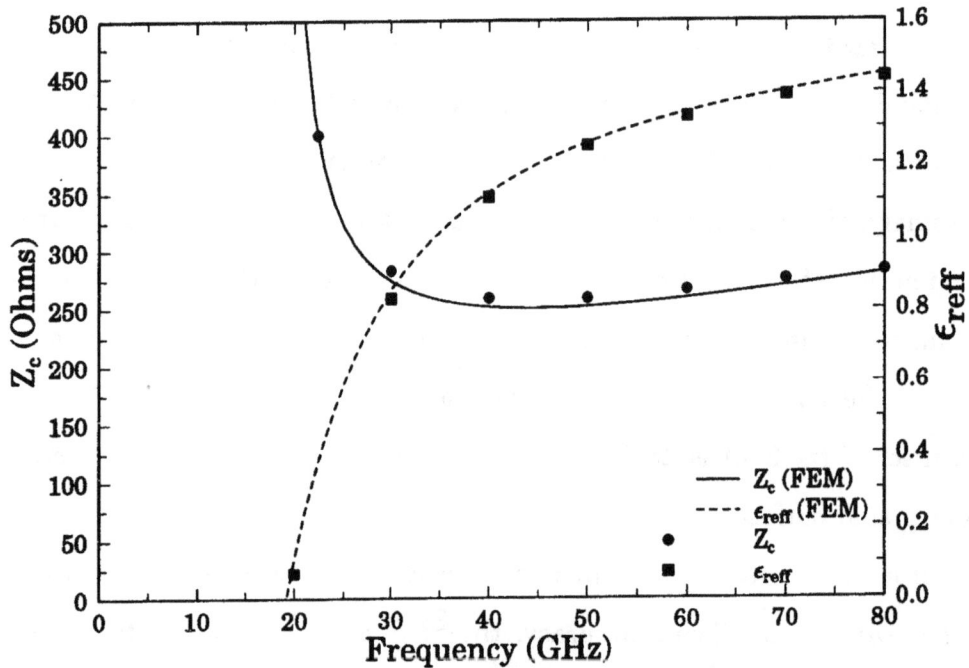

Fig. 2.4: Characteristic impedance versus frequency for a unilateral finline; $a = 2b = 4.7752$ mm, $s = 0.127$ mm, $h = 2.3876$ mm, $d = 0.47752$ mm, $\epsilon_r = 3.8$. The markers represent data extracted from a paper by Mansour *et al.*.

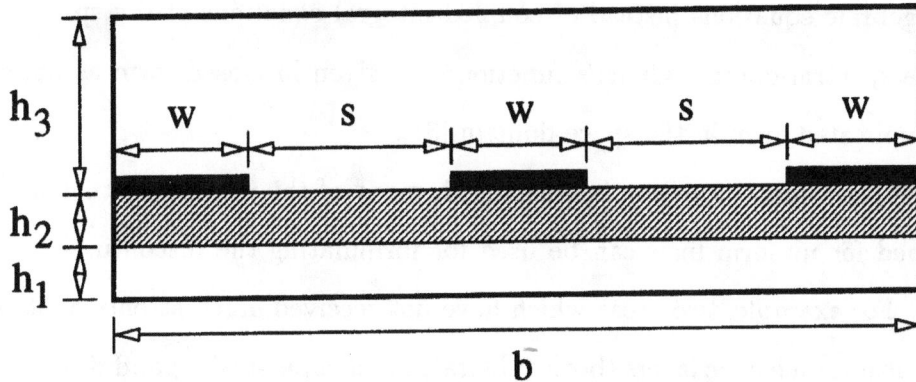

Fig. 2.5: Geometry of a suspended coplanar waveguide. Dimensions: $a = 7.112$ mm, $b = 3.556$ mm, $h_1/a = 0.4$, $h_2/a = 0.1$, $h_3/a = 0.5$, $w = s = b/5$.

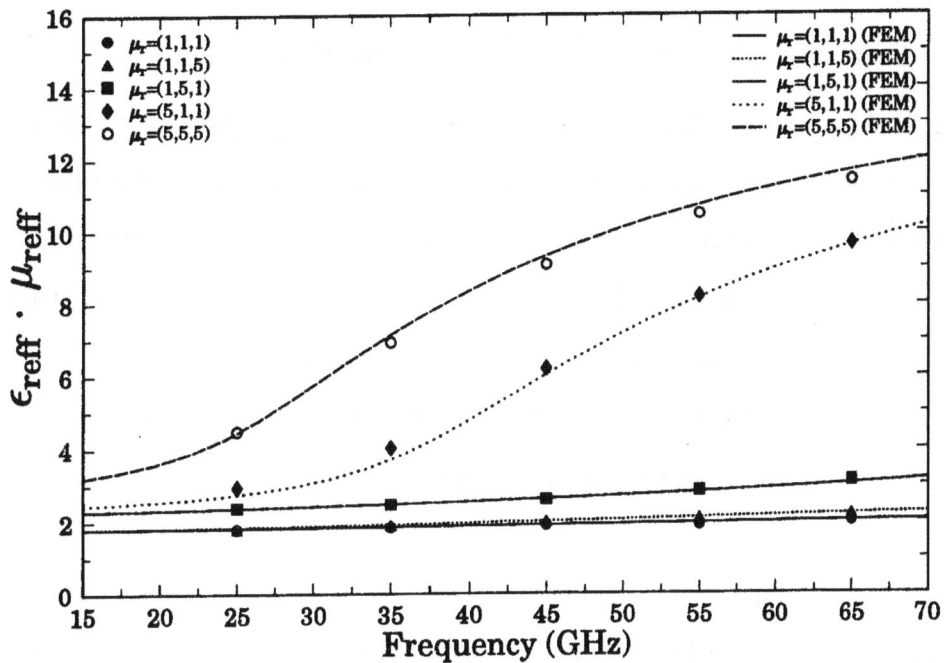

Fig. 2.6: Dispersion curves for the dominant mode of a suspended coplanar waveguide. The markers represent data extracted from a paper by Mazé-Merceur *et al.*.

conductors and discontinuities are all rigorously accounted for in the computations. Formulating the problem in the spectral domain is easier since the equations are coupled algebraic equations instead of coupled integral equations of the space domain, and the spectral-domain Green's functions are given in closed form as opposed to the complicated form in the space domain [31].

Since the Green's function is unchanged, many of the techniques that have been developed for uniform lines can be used for formulating the discontinuity problem as well. For example, two areas which have not received much attention, structures with multiple dielectric layers (both substrates and superstrates) and discontinuities on different dielectric interfaces, can easily be included in the analysis using a simple recursion formula [32] in combination with the Spectral-Domain Immitance approach [33] to compute the appropriate Green's function.

2.2.1 Novel method for the design of low-coupling structures

Using SDA, a novel method was discovered for the compensation of the even and odd mode phase velocities, using an additional substrate instead of a superstrate in microstrip and coplanar waveguide (CPW) structures [34],[35]. This method for designing low-coupling and low-distortion structures is relatively invariant with respect to frequency, and it does not hinder the placement of circuit elements. The key to the method is to choose a lower substrate with a relative dielectric constant that is much smaller than that of the upper substrate. This has been accomplished for both microstrip and CPW type structures.

As a first example of this method for coupled microstrip lines, the low-coupling height ratios for a multi-layer substrate coupled microstrip line are plotted in Fig. 2.7 as a function of the lower substrate dielectric constant. There are three different regions of interest when considering such a design which correspond to two different low-coupling point regions and a minimum coupling region [36]. For the case shown in Fig. 2.7, it is evident that there are two different height ratio combinations for which low-coupling will occur when the lower substrate dielectric constant is less than ≈ 4. However, for lower dielectric constant values of $4.0 \leq \epsilon_{r1} \leq 9.5$, there is only one height ratio value that minimizes the difference between even- and odd-mode phase velocities. The value of ϵ_{r1} at which the split occurs, i.e. $\epsilon_{r1} = 4.0$, corresponds to the maximum value for which the even- and odd-mode phase velocities can be exactly equalized.

The concept of equalizing even- and odd-mode phase velocities can also be applied to 3-slot edge-coupled CPW structures [35]. As previously done for coupled microstrips, the low-coupling height ratios for a multi-layer substrate edge-coupled CPW line are plotted in Fig. 2.8 as a function of lower substrate dielectric constant. The same three different height combination regions are also seen when considering an edge-coupled multi-layer substrate design that were seen in the microstrip case, corresponding to two different low-coupling point regions and a minimum coupling region. The lower portion of Fig. 2.8 corresponds to the first modal equalization height ratio, or first low-coupling height ratio, while the upper portion corresponds to the second low-coupling height ratio. As the lower substrate dielectric constant is varied for values $1.0 \leq \epsilon_{rL1} \leq 7.5$, both low-coupling points move closer to each other and eventually meet at a single substrate height combination. For values of $\epsilon_{rL1} > 7.5$, there is no substrate height combination for which the even- and odd-modes can be equalized. Since, this upper range of values does not lead to modal equalization, a value of $\epsilon_{rL1} \approx 7.5$ is considered an upper limit for the dielectric constant of the lower substrate, similar to that seen in the coupled microstrip case. For dielectric constants above this maximum value, only a single height ratio can be found that minimizes the difference between the phase velocities of each mode, as shown in Fig. 2.8. However, to guarantee at least two modal equalization points, the dielectric constant of the lower substrate, ϵ_{rL1}, should be chosen such that it is much smaller than that of the upper substrate, ϵ_{rL2}.

In practical circuit designs of low-coupling structures using multi-layer substrates,

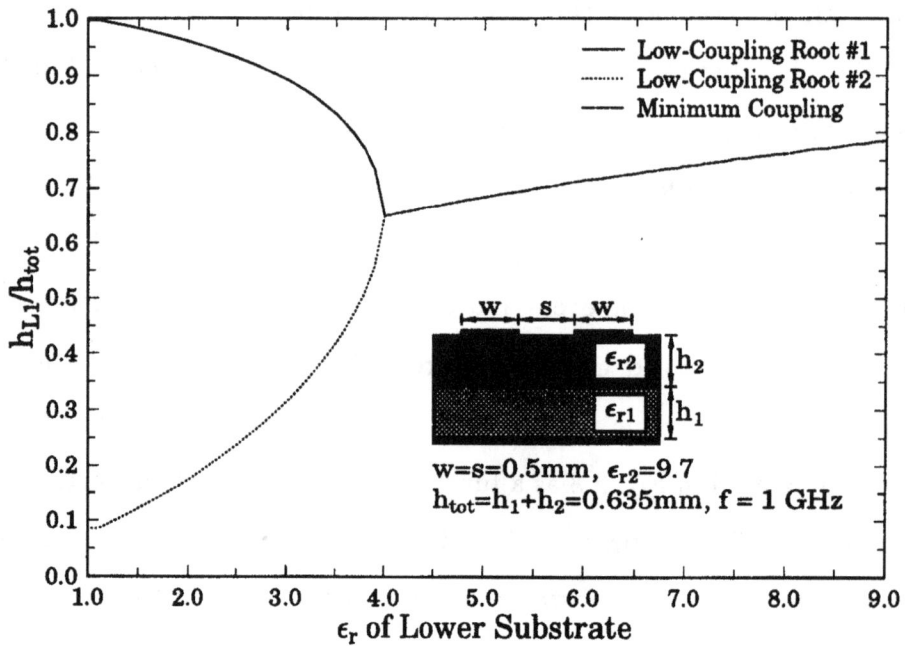

Fig. 2.7: Low-coupling height ratio points as a function of lower substrate dielectric constant for a compensated coupled microstrip structure.

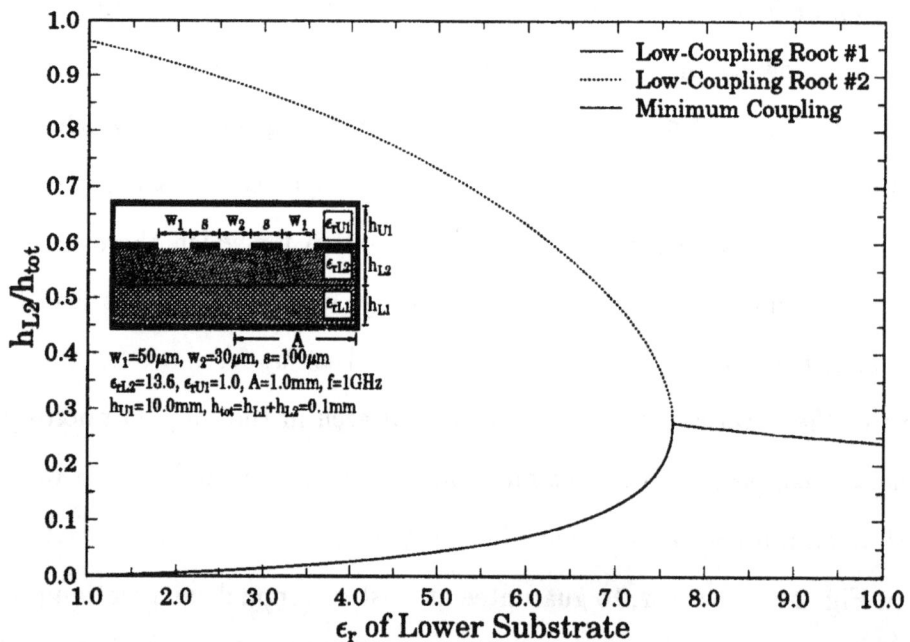

Fig. 2.8: Low-coupling height ratio points as a function of lower substrate dielectric constant for a compensated edge-coupled CPW structure.

the choice of substrate dielectric constants and heights is extremely important. To achieve low-coupling in a multi-layer coupled microstrip or edge-coupled CPW structure, the upper substrate dielectric constant must be greater than the lower substrate dielectric constant, as previously noted. It was demonstrated that there are two low-coupling height ratios for such structures. If the dielectric constant of the upper substrate is close in value to the lower substrate, the low-coupling height ratios tend to vary as a function of frequency. However, if the dielectric constant of the upper substrate is much greater than that of the lower substrate, the low-coupling height ratios remain essentially constant as a function of frequency, which is important in the design of broadband low-coupling and low pulse distortion structures [34],[35]. Typically, to ensure at least two different low-coupling points that are relatively constant as a function of frequency, the ratio of upper substrate dielectric constant to lower substrate dielectric constant must be at least 2:1.

The choice of low-coupling height ratio is also important. Depending on which height ratio is chosen, different combinations of thick and thin substrates are needed. Typically, the first low-coupling height ratio requires a very thin upper substrate above a thick lower substrate while the second low-coupling height ratio uses a thick upper substrate above a very thin lower substrate. The combination of substrates is chosen based upon the circuit application and typical materials used. For example, a circuit using primarily gallium arsenide ($\epsilon_r = 13.6$) as a substrate would most likely use the second low-coupling height ratio with a very thin lower substrate of smaller dielectric constant (such as air $\epsilon_r = 1.0$ or duroid $\epsilon_r = 2.2$) inserted below the gallium arsenide substrate. However, a circuit using primarily duroid ($\epsilon_r = 2.2$) as a substrate may use the first low-coupling height ratio with a very thin upper substrate of high dielectric constant inserted above a lower duroid substrate.

The effectiveness of this modal equalization technique in reducing pulse distortion and crosstalk is demonstrated for coupled microstrip lines. Two examples are considered, one using a single substrate layer and another using a multi-layer low-coupling design as discussed previously. Fig. 2.9 depicts the time domain response of a Gaussian shaped pulse applied at the input port of a single layer substrate coupled microstrip line after traveling a length of $l = 50$ mm. A reference undistorted pulse is shown for comparison purposes. The pulse seen at the direct port appears distorted in terms of amplitude reduction and pulse widening due to the difference in even- and odd-mode phase velocities and dispersion in the line. There is also a significant signal response seen at the crossover port due to the difference in even- and odd-mode phase velocities which degrades the intended signal response input at the direct port. The transient crosstalk seen at the crossover port is generally an undesirable result, especially in high-speed, high-density interconnects which require small inter-line spacings.

Next, a low-coupling height ratio, as shown in Fig. 2.7, is chosen to equalize the even- and odd-mode phase velocities using multi-layer substrates for a coupled microstrip structure. Fig. 2.10 shows the transient signal response of a Gaussian shaped pulse applied at the input port of a multi-layer substrate coupled microstrip line after traveling a length of $l = 50$ mm. In contrast to the previous single layer case, the pulse at the direct port has undergone almost no degradation of amplitude nor has it widened noticeably compared to the undistorted signal. The signal response seen at the crossover port has also been reduced, with an amplitude of only 15 % of the amplitude of the original pulse. This is a significant improvement in isolation between direct and crossover ports as compared to the previous single layer substrate coupled microstrip case.

To demonstrate the effectiveness of this method for CPWs, two different cases using an edge-coupled CPW coupler are considered. The first example uses a single substrate of gallium arsenide ($\epsilon_r = 13.6$). The effective dielectric constant (ϵ_{reff}) of the even and odd modes of such a structure is plotted in Fig. 2.11 as a function of frequency. Note that the even and odd mode effective dielectric constants have different values over all frequencies. This corresponds to each mode traveling at

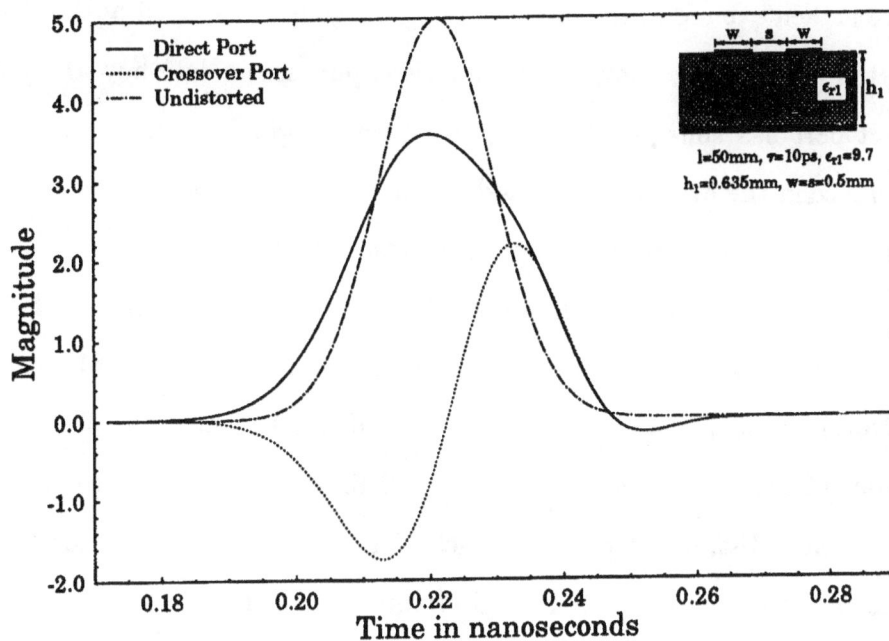

Fig. 2.9: Pulse distortion and crosstalk for two coupled microstrips using a single substrate ($l = 50$mm).

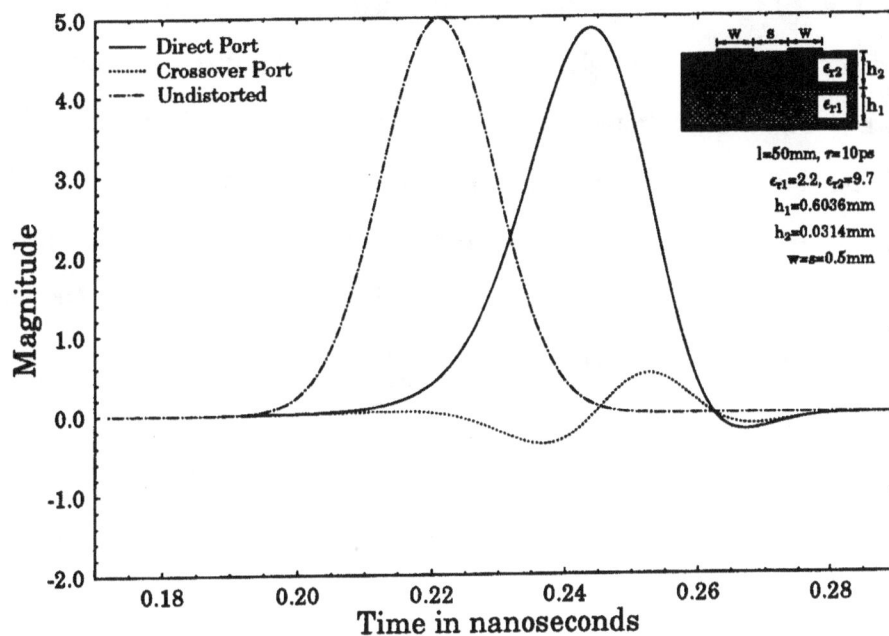

Fig. 2.10: Pulse distortion and crosstalk for two coupled microstrips using multi-layer substrates ($l = 50$mm).

different velocities at a given frequency. Fig. 2.12 demonstrates what happens if this structure is then excited at the direct port with a gaussian shaped pulse. A reference undistorted pulse is also shown for comparison purposes. Note how the pulse from the direct port has suffered distortion and even "ringing" on the front end due to the difference in the even and odd mode phase velocities. The crossover port of this structure also shows significant coupling effects.

The second case attempts to minimize the distortion and coupling in the previous structure using two substrate layers with gallium arsenide ($\epsilon_r = 13.6$) above a layer of RT/Duroid 5880 ($\epsilon_r = 2.2$). The effective dielectric constant of the even and odd modes of such a line is plotted in Fig. 2.13 as a function of frequency. The heights of the substrates are chosen such that the even and odd mode effective dielectric constants are equal at a single frequency. Although this guarantees that the even and odd mode phase velocities are equal at only one frequency and no others, the relative difference between the even and odd modes is minimized as shown in Fig. 2.13 compared to the previous single substrate case in Fig. 2.11. Using this modified structure, the pulse responses, at the same distance used in Fig. 2.12, are shown in Fig. 2.14. A reference undistorted pulse is also shown. The direct port response has shifted only slightly relative to the undistorted pulse, and has decreased in amplitude by 3 %. At the crossover port, the coupling amplitude has increased to almost 8 % of the undistorted pulse. A dramatic increase in isolation between direct and crossover ports is achieved using this new two substrate low-coupling design compared to using only a single substrate. This novel design technique enables the circuit designer to realize low-distortion, low-coupling structures that effectively decreases pulse degrading effects due to the difference between even and odd mode phase velocities.

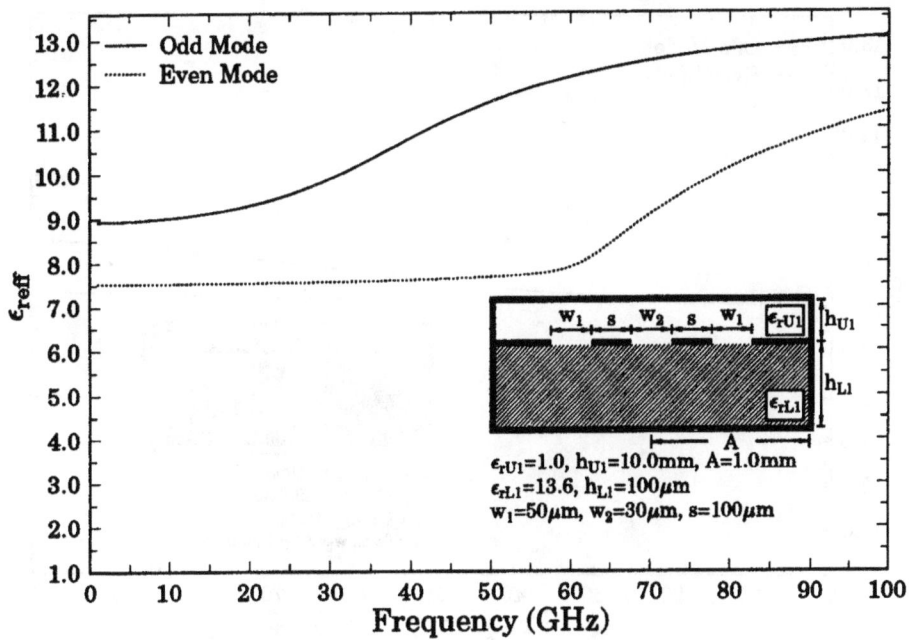

Fig. 2.11: ϵ_{reff} as a function of frequency for an edge-coupled CPW using a single substrate.

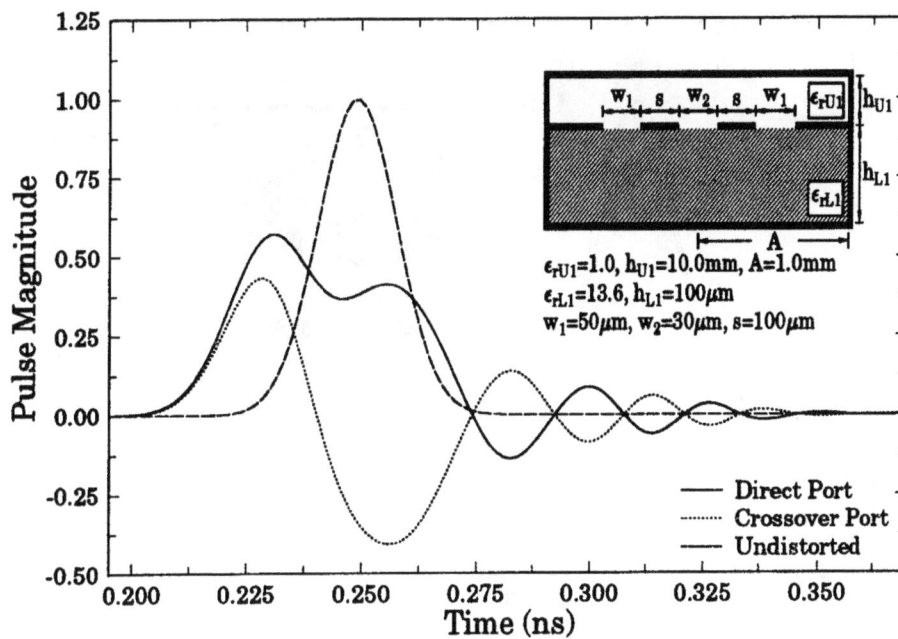

Fig. 2.12: Pulse distortion and crosstalk for an edge-coupled CPW using a single substrate ($l = 25$mm).

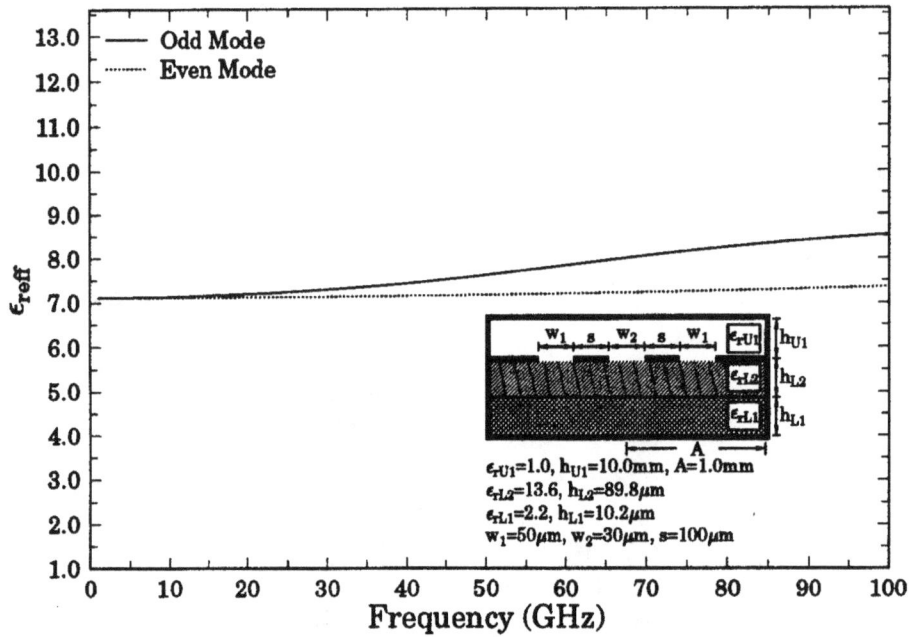

Fig. 2.13: ϵ_{reff} as a function of frequency for an edge-coupled CPW using two substrates.

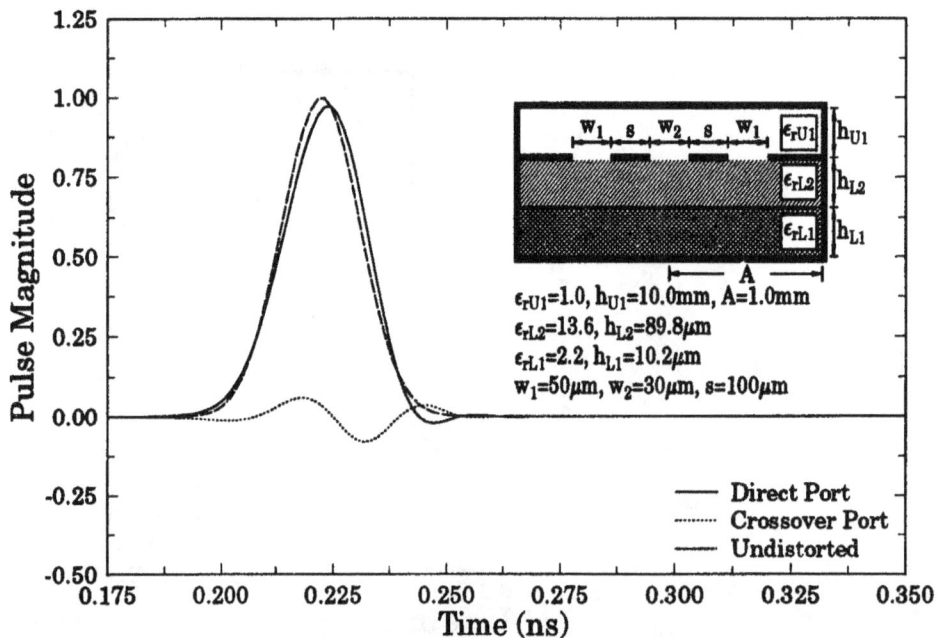

Fig. 2.14: Pulse distortion and crosstalk for an edge-coupled CPW using two substrates ($l = 25$mm).

2.2.2 Improved performance of microstrip couplers

A. Introduction

Fabrication techniques limit the level of coupling that can be obtained with edge-coupled microstrip lines while differences in the phase velocities degrade coupler performance. Tight couplings are difficult to achieve in edge-coupled microstrip lines due to the small spacing required. Small spacings and line widths also increase the conductor loss in the coupler. In addition, coupled microstrips on single substrate layers have different even and odd mode phase velocities, which degrades the performance of the coupler. Optimum coupler design requires equalized phase velocities for wide range of coupling values, even for tight couplings, using practical line widths and conductor spacings. The design must also be realizeable, using techniques and materials that are commercially available. All of these requirements can be met using substrate compensation to design edge-coupled microstrip directional couplers.

One technique that can be used to equalize the even and odd mode phase velocities is capacitive compensation, where the capacitance of the odd mode is increased by including lumped capacitors at either end of the coupling section [37]. While this provides better isolation, it also makes it more difficult and expensive to fabircate and does not improve the coupling performance of the circuit. Another technique that can be used is to place a dielectric layer on top of the coupled section [38], [39]. Although this technique allows the phase velocities to be equalized, care must be taken to insure that there is not an air gap between the substrate and the overlay dielectric [38]. In addition, in order to maintain the proper impedances and coupling, the lines widths and spacings must be decreased, increasing the conductor loss and making the circuit more difficult to fabricate.

However, the phase velocities of the even and odd modes can also be equalized using an additional substrate layer that has a dielectric constant that is much less that of the substrate layer above it [34]. This technique, referred to as *substrate compensation*, has been shown to significantly decrease the transient, or forward, coupling in coupled microstrip lines by equalizing the even and odd mode phase velocities [34]. However, while the forward coupling, which is dependent on the phase velocities, is reduced, the steady-state, or backward, coupling is actually enhanced. Furthermore, the center conductor spacing required for a given input impedance and coupling coefficient is larger for a substrate compensated structure than for one using a single substrate of either material. This allows tighter couplings in edge-coupled configurations and decreases the conductor loss. Finally, substrate materials are commercially available with the proper electrical properties and matched thermal coefficients of expansion to allow the realization of this type of high performance directional coupler.

B. Two-substrate coupled lines

The geometry for substrate-compensated microstrip lines is shown in Fig. 2.15. The key to substrate compensation is to choose a lower substrate with a dielectric constant, ϵ_{r1}, that is much less than the dielectric constant of the upper substrate, ϵ_{r2}. Typically, ϵ_{r1} should be about one-third or one-half of the value of ϵ_{r2}. Choosing the proper combination of substrate heights and center conductor geometry will give the desired coupling coefficient and input impedance with the phase velocities equalized. The Spectral Domain Approach is used to compute the effective dielectric constants of the even and odd mode [34], [36] and a non-linear optimizer is used to determine the values of the substrate heights and center conductor geometries that satisfy the design criteria. All of the cases presented here are designed for a 50Ω input impedance.

Results using this technique for the center conductor width and spacing are shown in Fig. 2.16 for suspended coupled microstrips on alumina in the quasi-static region. The results are compared with those given by commercially available design software, LineCalc™, Version 1.3. In both figures, the total substrate height, $h_{tot} = h_1 + h_2$ is held constant while the ratio of h_1/h_{tot} is varied. When the height ratio is 0.0, the

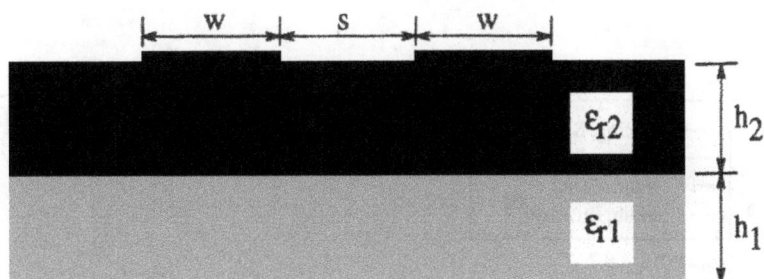

Fig. 2.15: Coupled microstrips with two substrate layers used for substrate-compensated directional couplers.

structure is a single substrate of alumina and when it is 1.0, it is an air line. For small values of the substrate height ratio, $h_1/h_{tot} < 0.5$, the results for the conductor width agree fairly well. For larger values of the substrate height ratio, the results from the SDA predict smaller values of the conductor width. The results for the center conductor spacing have a much larger disagreement. The results from LineCalc^{TM} have a maximum center conductor spacing for height ratio near 0.3, while for the SDA data the maximum conductor spacing occurs at a value of the height ratio near 0.75. In addition to the large difference in the values of the center condcutor spacing, the overall behavior of the two graphs is much different.

In Fig. 2.17, the ϵ_{reff} of the even and odd modes are shown as a function of the height ratio for the suspended microstrip structure in Fig. 2.16. The results for the even mode from LineCalc^{TM} are slightly higher than the full-wave results, but show the same general behavior. On the other hand, the results for the odd mode for LineCalc^{TM} begin to deviate significantly at a height ratio of 0.5. As the height ratio increases above 0.5, the odd mode ϵ_{reff} does not approach a value of 1.0, which is the value for an air-filled line. However, both techniques predict the crossing of the ϵ_{reff} for the even and the odd modes near a height ratio of 0.15. Near this value of the height ratio, the even and odd modes have nearly the same phase velocities. Thus edge-coupled microstrip directional couplers designed on suspended substrate with these dimensions will have increased the directivity due to the matching of the even and odd mode phase velocities.

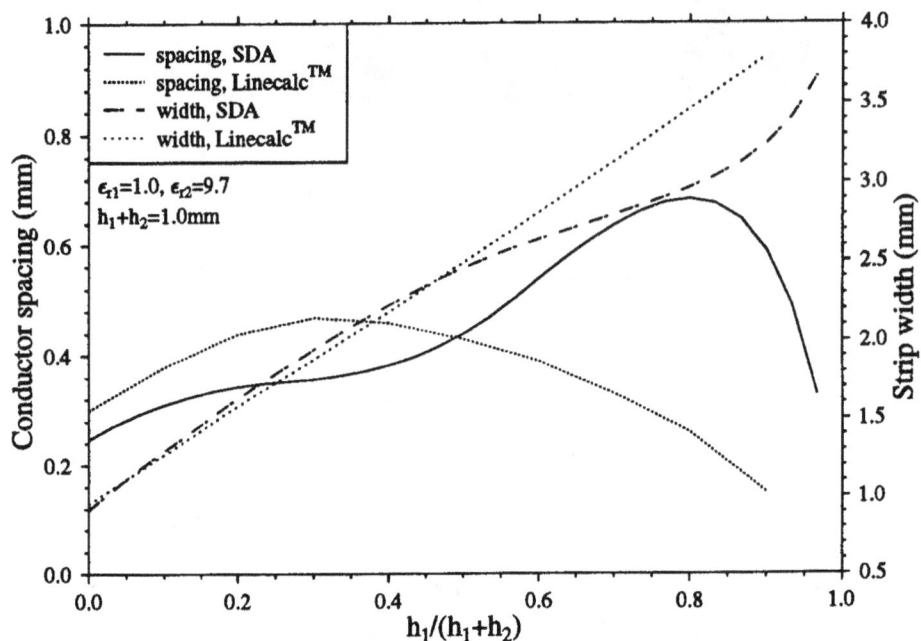

Fig. 2.16: Conductor spacing and width versus the substrate height ratio for suspened coupled microstrips with 50Ω input impedance and 10 dB coupling.

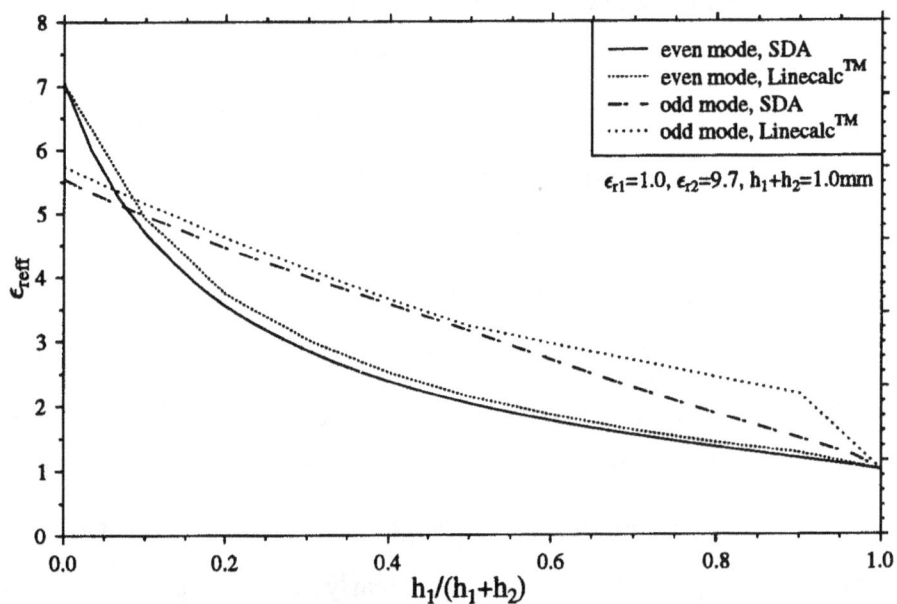

Fig. 2.17: Effective dielectric constant of the even and odd modes as a function of the substrate height ratio for suspended coupled microstrips with 50Ω input impedance and 10 dB coupling.

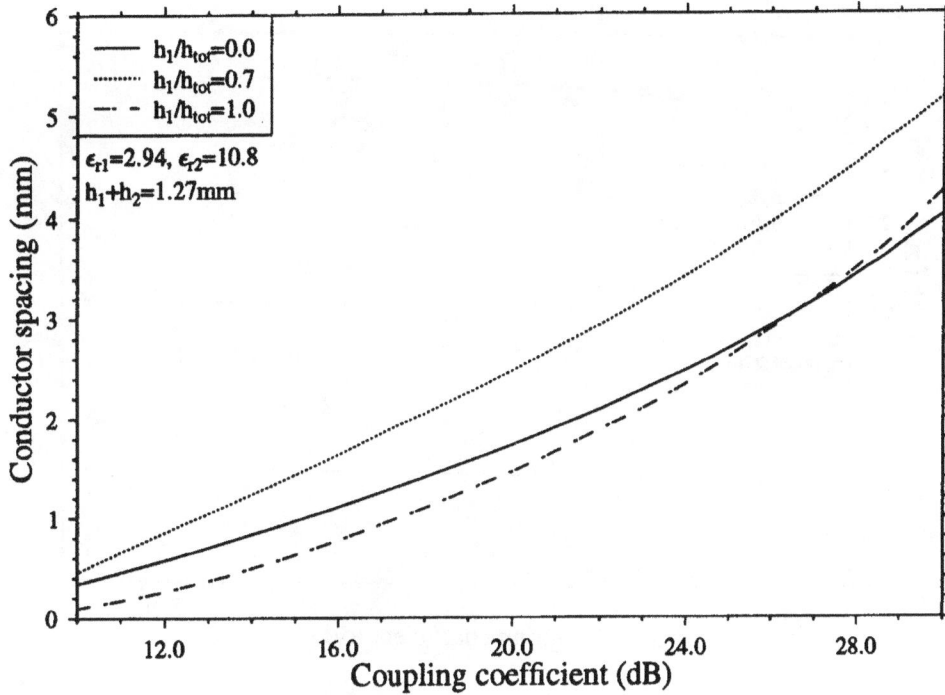

Fig. 2.18: Center conductor spacing as a function of the coupling coefficient for three substrate height ratios.

In the next three figures, Figs 2.18 through 2.20, the spacing, center conductor width, and $\epsilon_{r\text{eff}}$ for coupled microstrips is shown for three different substrate heigth ratios, 0.0, 0.7, and 1.0 as a function of the coupling coefficient. The upper substrate is RT/Duroid 6010, $\epsilon_r = 10.8$, and the lower substrate is RT/Duroid 6002, $\epsilon_r = 2.94$, both of which are commercially available and have matched thermal characteristics. When the height ratio is zero, the structure is a single substrate of 6010 and when it is 1.0, it is a single substrate structure of 6002. In Fig. 2.18 the center conductor spacing is shown for values of the coupling coefficient from 10dB to 30dB. For all values of the coupling coefficient, the spacing required for the compensated substrate is greater than for either of the single substrate cases. On the other hand, the center conductor width, shown in Fig. 2.19, for the compensated substrate is between the widths required for the two single substrate cases.

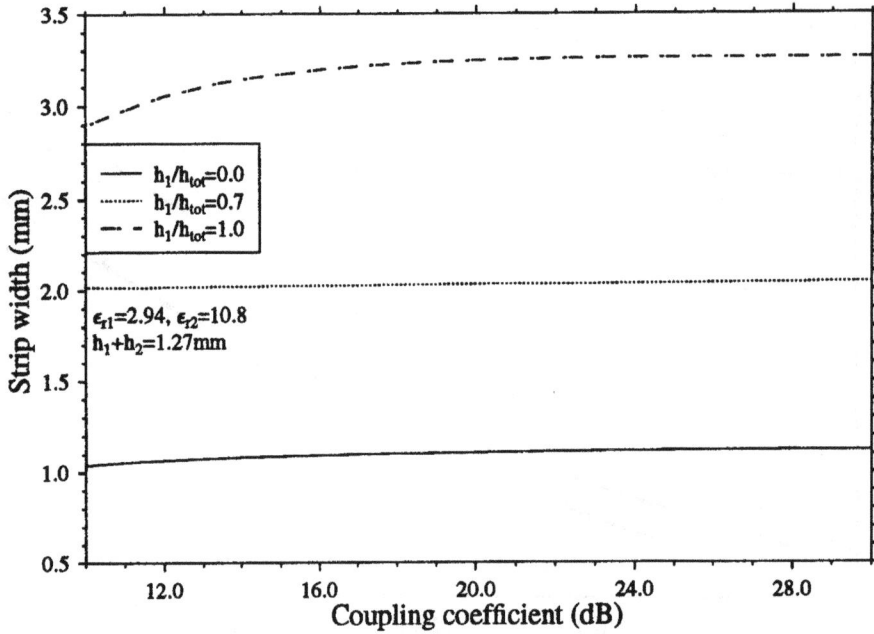

Fig. 2.19: Width of the center conductors in mm as a function of the coupling coefficient for three substrate height ratios.

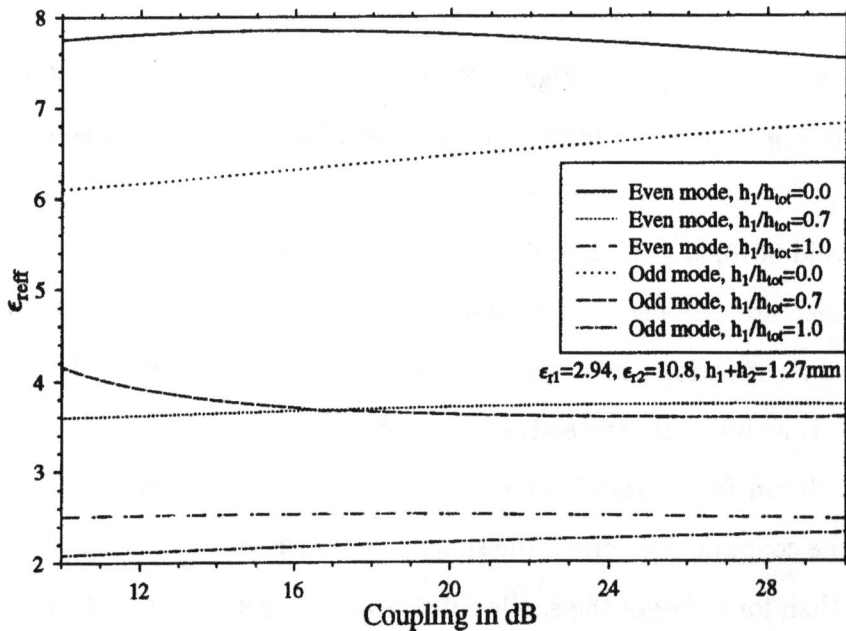

Fig. 2.20: Even and odd mode ϵ_{reff} as a function of the coupling coefficient for three substrate height ratios.

The compensated substrate also gives a better match between the even and odd mode $\epsilon_{r\text{eff}}$, as shown in Fig. 2.20. Both single substrate structures show a large difference between the modal $\epsilon_{r\text{eff}}$'s for all values of the coupling coefficient, increasing as the coefficient is decreased. However, the compensated substrate has little difference between the even and odd mode $\epsilon_{r\text{eff}}$, except at very tight coupling. Note that these results were produced using a fixed substrate height ratio. It it possible to design the coupled lines such that the difference in the modal $\epsilon_{r\text{eff}}$'s is minimized by choosing the proper substrate height ratio.

C. Conclusion

Edge-coupled microstrip directional couplers are limited by the conductor spacing and widths that can be realized in commercial processes. Differences in the even and odd mode phase velocities also degrade the performance of the coupler. However, substrate compensation can be used effectively to equalize the phase velocities on edge-coupled microstrip directional couplers. In addition, the center conductor spacing requied for a given input impedance and coupling is increased, allowing easier fabrication. This combination of equalized phase velocities and wider center conductor spacings allows the design of microstrip directional couplers with tighter couplings and better performance.

2.2.3 Wide-band coupling reduction

A. Introduction

Ultrafast clock speeds and short rise times of digital signals give them a very large bandwidth. With clock speeds in the Gigabit/second range and rise times on the order of 10 picoseconds, these signals will have significant frequency components in the tens to hundreds of GHz. This wide frequency range requires a full-wave analysis, rather than a a quasi-static one, to accurately predict the distortion and coupling of high-speed digital signals. High speed circuits also have small inter-line spacing to increase packing density, which in turn increases coupling distortion and crosstalk. Successful design of these high-speed circuits requires accurate prediction of pulse distortion and coupling, as well as a technique to reduce coupling over a very wide band of frequencies.

Many different techniques have been used to reduce coupling in high-speed digital circuits. One approach is to place an extra line that is grounded at both ends between adjacent lines[40, 41]. Although this method can provide some reduction in crosstalk, it increases fabrication costs, increases inter-line spacing, and requires via holes. In addition, while the direct crosstalk is reduced, the return voltage can increase the near-end crosstalk. Another technique that can be used is substrate compensation which reduces crosstalk is through the choice of the electrical parameters of a multi-layer substrate[34]. This technique allows very small inter-line spacings, less than one conductor width, while reducing coupling over a very wide range of frequency. For the case on two adjacent conductors, the crosstalk and coupling distortion can be essentially eliminated. While it is not always possible to completely eliminate crosstalk for multi-conductor cases, it is usually possible to obtain a significant reduction in the level of crosstalk through substrate compensation.

In this report, distortion and coupling of wide-band signals on integrated circuit lines is analyzed using a full-wave approach to obtain frequency-domain data, and a Fourier transform approach to obtain the time-domain results. The effect of additional lines on the distortion and crosstalk of multi-line structures is examined, showing how crosstalk can be significantly reduced for a wide range of conductor configurations. Variation of crosstalk and distortion for different pulse widths is also investigated to illustrate the effect of dispersion for multi-conductor lines.

B. Results

The geometry for a multi-layer, multi-conductor structure is shown in Fig. 2.21. Only one conductor is shown, with a width w_i located at x_i, although any finite number

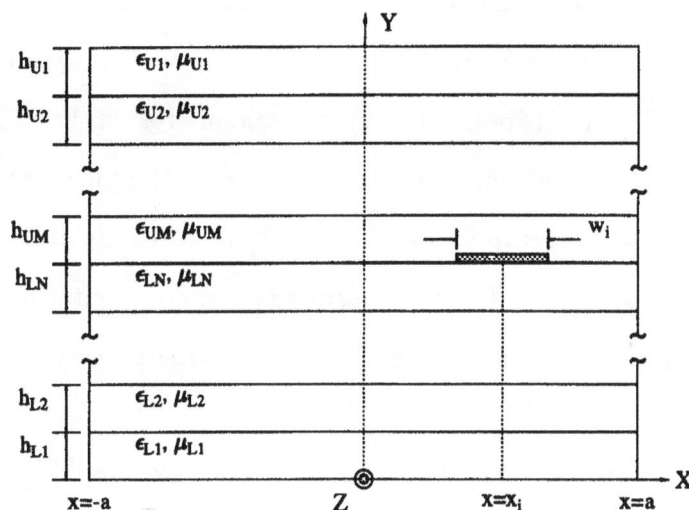

Fig. 2.21: Geometry of a multi-layer, multi-conductor interconnect.

can be included in the analysis. The interline spacing, s_i, is measured between the adjacent edges of conductors. All of the cases considered here are open laterally, $a \rightarrow \infty$, with no cover layer, $h_{U1} \rightarrow \infty$.

For all of the following results, a ramped pulse is used with a time domain response given by:

$$v(t) = \begin{cases} 0 & \text{for} \quad |t| > \tau + \tau_r \\ 1 - \dfrac{|t| - \tau}{\tau_r} & \text{for} \quad \tau \leq |t| \leq \tau + \tau_r \\ 1 & \text{for} \quad |t| \leq \tau \end{cases}$$

where τ_r is the rise time and τ is the pulse duration. Only symmetric conductor configurations are considered and the lines are numbered consecutively from one side to the other, i.e. for the four conductor case, line 1 and line 4 are on the outside, with lines 2 and 3 in the center.

To design a substrate-compensated, low-coupling structure, the substrates materials and heights need to be chosen such that the maximum level of crosstalk on the adjacent lines is minimized. For the single layer case, the lines closest to the excited line have the highest level of crosstalk. However, for the multi-layer case, the crosstalk level on the closest lines may be less than on those further away.

The maximum crosstalk on adjacent lines as a function of substrate height ratio, $h_{L1}/(h_{L2} + h_{L1})$ is shown in Fig. 2.22 for two, three and four symmetric microstrip lines and in Fig. 2.23 for two and five symmetric microstrip lines. The maximum crosstalk is calculated using the quasi-static values of the phase constants and the current matrix for a ramped pulse with $\tau = 100$ picoseconds and $\tau_r = 10$ picoseconds at a distance of 25 mm. Only the results for unique excitations are shown, for example, in the four conductor case exciting line 4 is equivalent to exciting line 1, since the structure is symmetric.

For the two-conductor case, shown in both Fig. 2.22 and Fig. 2.23, there are two values of the height ratio which eliminates the crosstalk on the adjacent line. Likewise, for the three-conductor case with the middle line (line 2) being excited, it is also possible to essentially eliminate the crosstalk on both of the adjacent lines. For this case, the behavior of the maximum crosstalk level is very close to the two-conductor case, except that the highest level of crosstalk is only about 0.36 compared with 0.5 for the two line case. These cases are similar because there is at most one conductor on either side of the excited line.

While it was possible to essentially eliminate the crosstalk for the 2 line case and the three line case with line 2 excited, for the four and five conductor cases, and the three conductor case with line 1 excited, there is a limit to the reduction of the crosstalk. The reason for this is that while the crosstalk to the nearest lines is greatly reduced, the crosstalk to lines farther away is not reduced as much and so the maximum crosstalk of all of the lines is larger than the crosstalk level on the nearest line. Thus, while it is usually possible to essentially eliminate the crosstalk to the nearest lines in a multi-conductor case, the overall crosstalk reduction is limited by the crosstalk to lines farther away.

Just as the 2-line case was similar to the 3-line case with line 2 excited, the crosstalk levels for 3 lines with line 1 excited and 4 lines with line 2 excited show very similar behavior, especially in the low-coupling regions. These cases are similar because there is at most 2 conductors on either side of the excited conductor. Thus it would be expected that the case of 4 lines with line 1 excited will give results similar to the following cases; 1) 5 lines, line 2 excited, 2) 6 lines, line 3 excited, and 3) 7 lines, line 4 (the middle line) excited, since these cases have at most three conductors on either side of the excited line. Indeed, the 5 line case with line 2 excited in Fig. 2.23 is similar to the 4 line case with line 1 excited in Fig. 2.22. Since additional lines, i.e. more than four on either side of the excited line, would have little effect on the crosstalk and coupling distortion, it is expected that the low-coupling regions shown in Figs. 2.22 and 2.23 would have reduced coupling for any number of lines of these widths and spacings on this substrate.

In the low coupling regions of Figs. 2.22 and 2.23, the crosstalk to the nearest lines is drastically reduced, even essentially eliminated, but the crosstalk on lines farther away is not reduced by as much and so these lines have the highest level of crosstalk. This effect can be seen in Fig. 2.24 where the maximum crosstalk as a function of the substrate height ratio on each of the individual lines is shown for the five conductor case with the outside line (line 1) being excited. The crosstalk level on the nearest line, line 2, is greatly reduced, from 0.42 to 0.04, for two different values of the substrate height ratio, as with the two conductor case. However, near those two values, the crosstalk on line 3 is much higher, and so the maximum crosstalk on all of the lines is not reduced as much. Furthermore, for height ratios between 0.35 and 0.85 the maximum crosstalk level occurs on line 4, two lines removed from the excited line. Thus for multi-conductor, multi-layer structures the highest crosstalk level is not necessarily on the nearest conductor. However, while it may not be possible to essentially eliminate the crosstalk on a multi-line structure with substrate compensation, it is still possible to get a substantial reduction in the crosstalk using substrate compensation. For the five line case, the maximum crosstalk level on the other lines can be a reduced by up to a factor of four.

Results for distortion and coupling of a ramped pulse on a four-conductor line

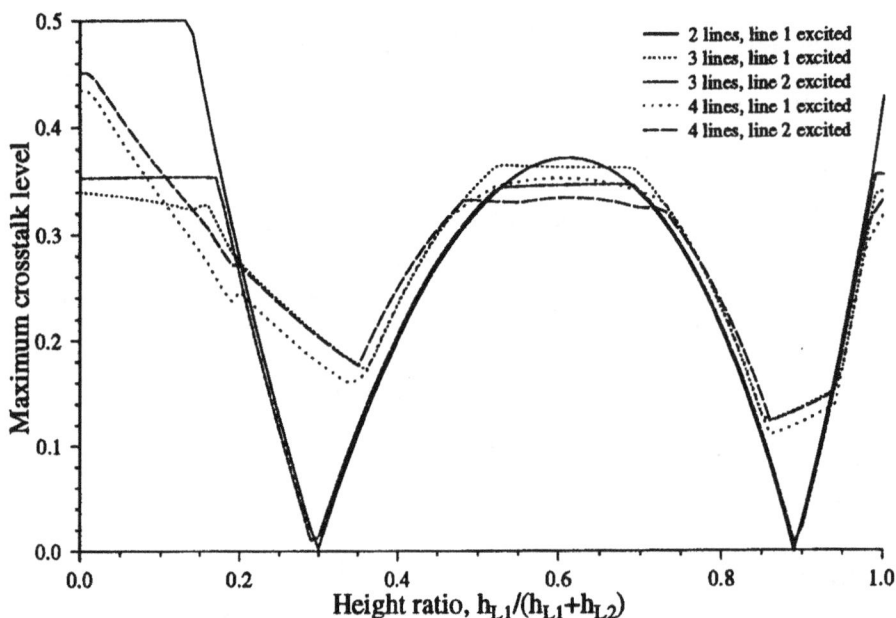

Fig. 2.22: Maximum crosstalk on a two-layer substrate for two, three, and four signal conductors as a function of the substrate height ratio, $h_{L1}/(h_{L1} + h_{L2})$ ($w = s = 0.6$ mm, $\epsilon_{rL1} = 2.94$, $\epsilon_{rL2} = 10.8$, $\epsilon_{rU1} = 1.0$, $h_{L1} + h_{L2} = 0.635$ mm, $h_{U1} \to \infty$).

are shown in Figs. 2.25 and 2.26 where $\tau = 100$ picoseconds and $\tau_r = 10$ picoseconds. For this example, line 1 is excited and the responses on all four lines are shown at a distance of $l = 50$ mm as a function of the normalized time, $(t - t_0)/\tau$, where t is time, $t_0 = (l/c)\sqrt{\epsilon_{reff(min)}(0)}$, c is the speed of light, and $\epsilon_{reff(min)}(0)$ is the minimum effective dielectric constant of the four modes in the quasistatic region. Fig. 2.25 is a single-layer substrate, while Fig. 2.26 is a two layer substrate designed to have low coupling through substrate compensation by choosing $h_{L1}/(h_{L1} + h_{L2}) = 0.875$.

In Fig. 2.25, the crosstalk on the adjacent line is almost 40% of the amplitude of the initial signal, representing a very high level of crosstalk. The highest level of crosstalk occurs on the nearest line, line 2. The pulse on line 1 has also broadened in time, due to coupling distortion. On the compenstated substrate, however, the maximum crosstalk on the adjacent lines is only 15% of the amplitude of the initial signal. The maximum crosstalk occurs on line 3, which is one removed from the

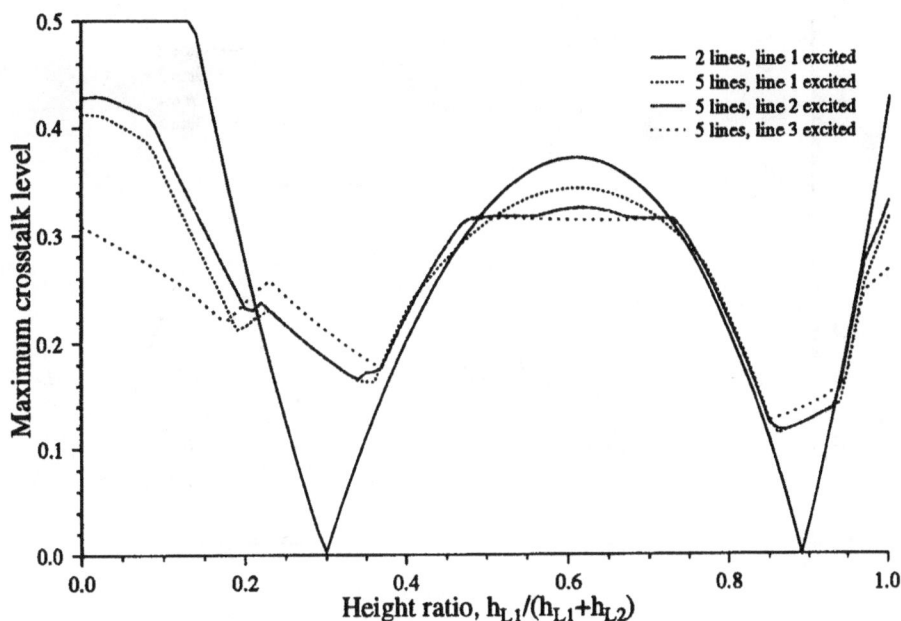

Fig. 2.23: Maximum crosstalk on a two-layer substrate for two and five signal conductors as a function of the substrate height ratio, $h_{L1}/(h_{L1} + h_{L2})$ with the same dimensions as in Fig. 2.22.

excited line. On the other hand, the response on line 2 is only 5% of the amplitude of the intial signal, the lowest of the three adjacent lines. Thus, this substrate configuration minimizes the coupling to the nearest lines, but only slightly decreases the coupling to lines farther away. However, the crosstalk has been substantially decreased using substrate compensation.

As the pulse duration, τ, or rise time, τ_r, decreases, the signal will suffer more degradation due to dispersion. Using the same structures as in Figs. 2.25 and Fig. 2.26, the time domain response on the four lines is shown in Figs. 2.27 and 2.28 for a ramped pulse with one-half the duration, $\tau = 50$ picoseconds, and the same rise time, $\tau_r = 10$ picoseconds. The pulse on the uncompensated substrate shows much more coupling distortion and pulse widening than the longer pulse in Fig. 2.25. The crosstalk on the adjacent lines is also much greater, due to the shorter pulse width. On the other hand, the pulse on the compensated substrate shows much

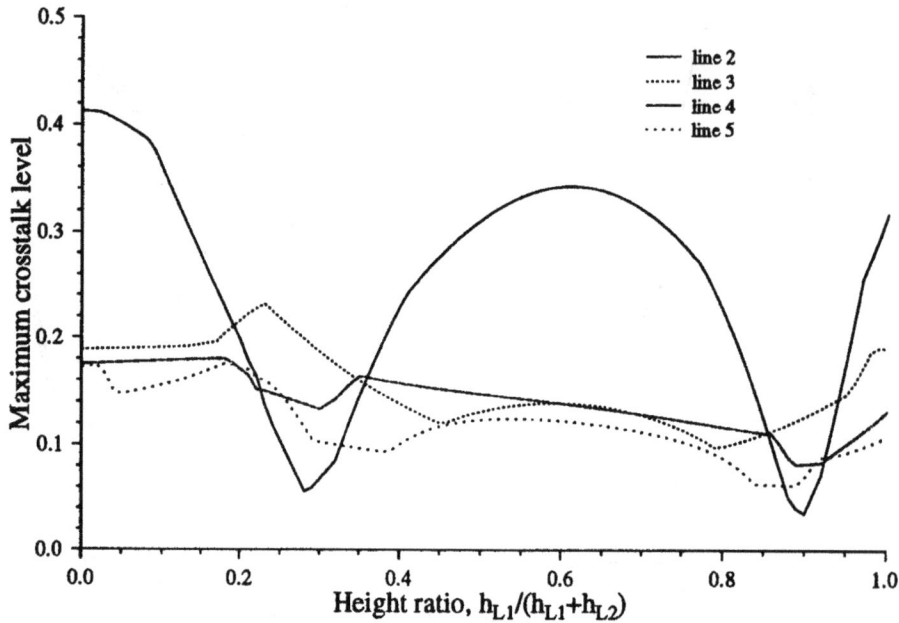

Fig. 2.24: Maximum crosstalk on the four adjacent lines of a two-layer structure with five signal conductors as a function of the substrate height ratio $h_{L1}/(h_{L1} + h_{L2})$ with the same dimensions as in Fig. 2.22.

less distortion due to coupling, and the crosstalk on the adjacent lines is much lower.

2.2.4 Closed-form expressions for the design of microstrip lines with two substrate layers

A. Introduction

As more microwave circuit applications move into the commercial field, considerations such as process yield, repeatability and manufacturability place additional demands on microwave design tools. While full-wave finite-element and moment-method *analysis* tools are now becoming available that provide very accurate results, *design* tools must trade some accuracy for speed, returning very good results in minutes, instead of hours. These design tools must be based on an accurate full-wave technique so that the designed element has the desired performance when used in the analysis tool. For example, a transmission line that is 50Ω in the design tool

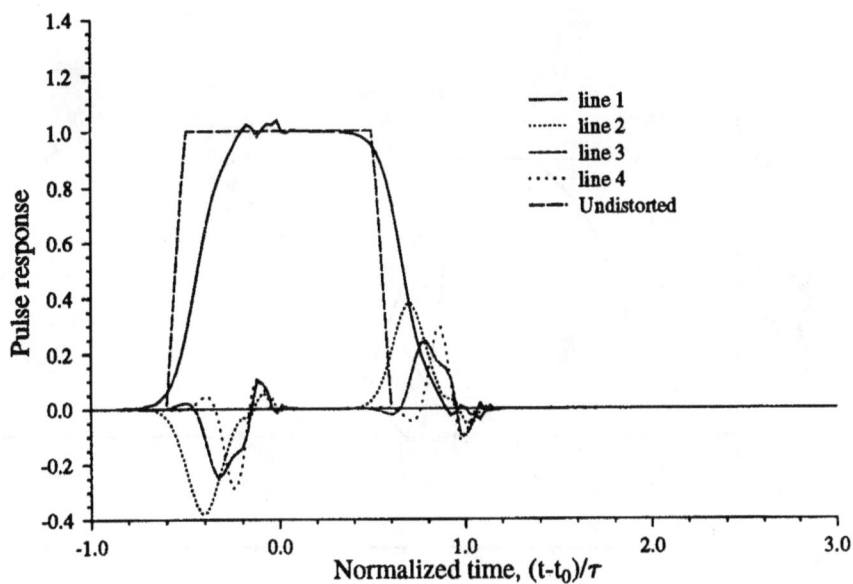

Fig. 2.25: Pulse distortion on four symmetric conductors at a distance $l = 50$ mm with line 1 excited ($w = s = 0.6$ mm, $\epsilon_{rL1} = 2.94$, $\epsilon_{rL2} = 10.2$, $\epsilon_{rU1} = 1.0$, $h_{L1} = 0.635$ mm, $h_{L2} = 0$ mm, $h_{U1} \to \infty$).

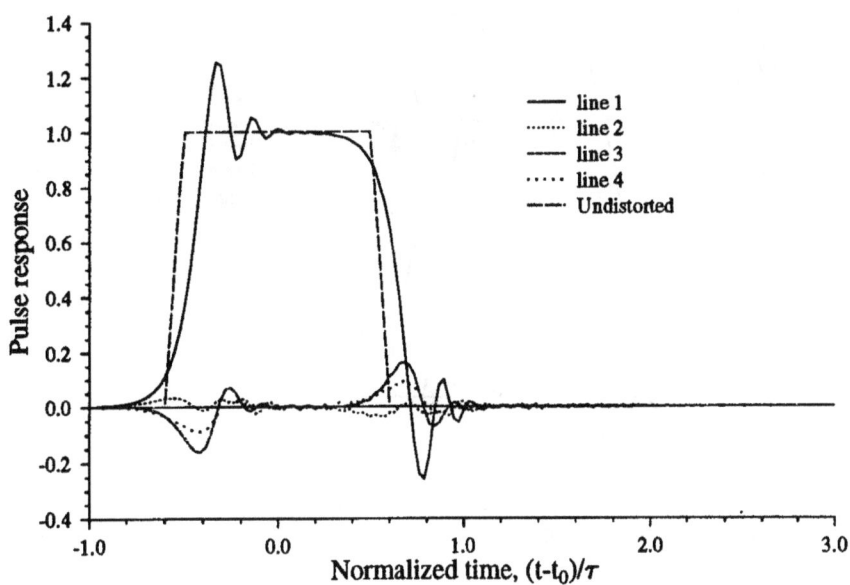

Fig. 2.26: Pulse distortion on four symmetric conductors at a distance $l = 50$ mm with line 1 excited ($w = s = 0.6$ mm, $\epsilon_{rL1} = 2.94$, $\epsilon_{rL2} = 10.8$, $\epsilon_{rU1} = 1.0$, $h_{L1} = 0.55562$ mm, $h_{L2} = 0.07938$ mm, $h_{U1} \to \infty$).

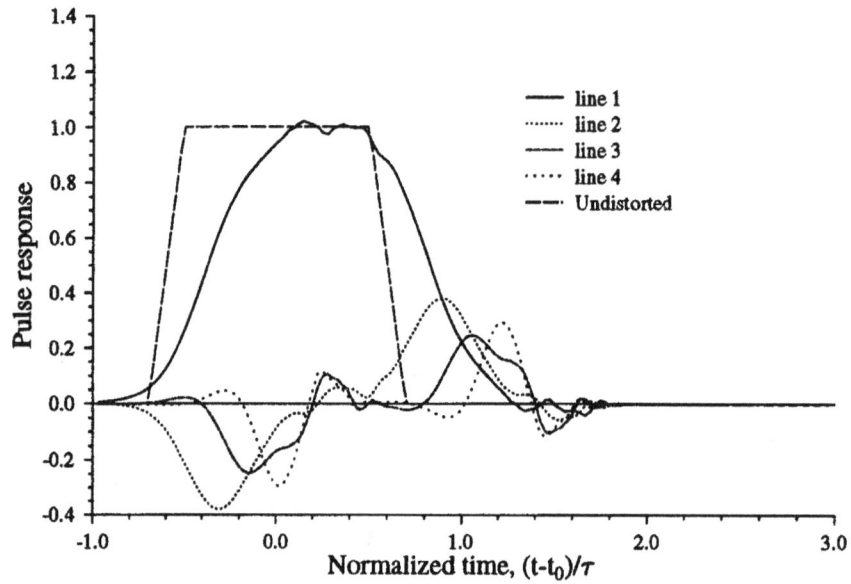

Fig. 2.27: Pulse distortion on four symmetric conductors at a distance $l = 50$ mm with line 1 excited and the same dimensions as in Fig. 2.25.

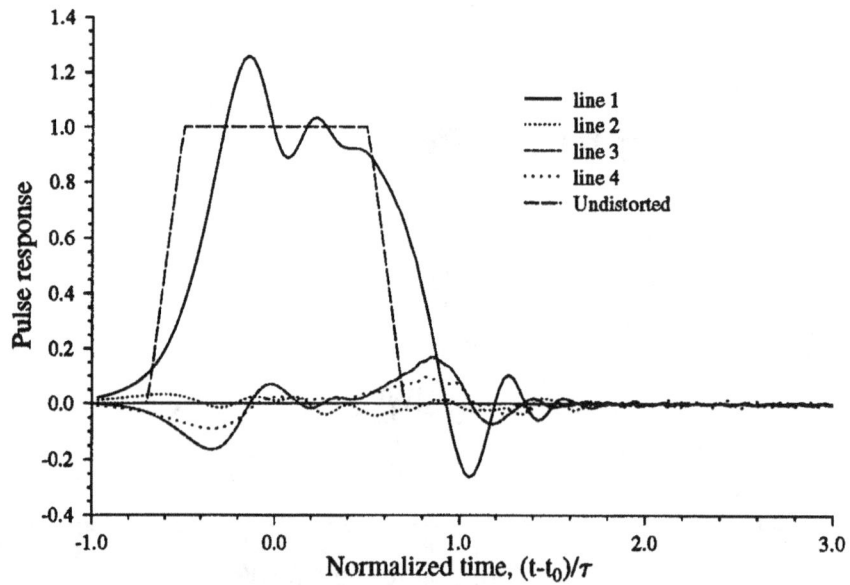

Fig. 2.28: Pulse distortion on four symmetric conductors at a distance $l = 50$ mm with line 1 excited and the same dimenstions as in Fig. 2.26.

should not evaluate to a 45Ω line on the analysis tool. To analyze the process yield, many similar cases must be analyzed to calculate the sensitivity of the design to process variation, emphasizing the need for very fast design tools. Complicating the microwave design problem is the increased use of multi-layer geometries, for which there are no accurate formulas available. Accurate and efficient design tools that can be used for multi-conductor, multi-layer geometries are required to achieve the goals of successful microwave design.

Accurate closed-form expressions for the transmission line parameters, ϵ_{reff} and Z_0, of open microstrips on a single substrate have been available for many years [42], [43]. Closed-form expressions for the transmission line parameters of coupled microstrips on suspended substrate have been made available more recently [44]. However, there are not any closed-form expressions for the parameters of open microstrips in more general structures, even for the simple single-conductor, two-layer substrate case. The goal of this research is to develop closed-form expressions for the design of open microstrip lines with two substrate layers.

B.   Analysis

The general trends of the parameters of a single substrate microstrip are well-known, e.g. wider center conductors have increased ϵ_{reff} and decreased impedance. However, for as few as two substrate layers, the general trends of the transmission line parameters become much more complicated. For example, if the dielectric constant of the upper substrate is much greater than that of the lower substrate, the ϵ_{reff} for the odd mode can be higher than that of the even mode [34], which is the reverse of the single layer case. Furthermore, increasing the width of the center conductor can actually decrease the effective dielectric constant for a single microstrip on a two-layer substrate. Thus it is important that any closed-form expression accurately take into account the unique behavior of multiple substrate structures.

The design tools proposed in this work have been generated using the Spectral

Domain Approach [45] to compute the transmission line parameters of single and coupled microstrips on two-layer substrates. A recurrence relation is used to compute the dyadic Green's function [34] and the expansion functions chosen are Chebyshev polynomials modifying the appropriate edge conditions [46]. Five longitudinal and four transverse current expansion functions are used to insure accurate results, with all calculations being done in double precision. The analysis is valid in the quasi-static region. The numerical results from the SDA are used to create curve-fitted formulas that are valid for a wide range of substrate parameters and conductor configurations. Wherever possible, the results have been compared to previously published data and other formulas to verify the accuracy of these new formulas.

The geometry of the two-substrate layer microstrip used in the present analysis is shown in Fig. 2.29. The total height of the substrate layers, $h_{tot} = h_1 + h_2$, is held constant and the design formula uses the following as design parameters:

- ratio of the width to total height, w/h_{tot}

- the height ratio, $h_r = h_1/h_{tot}$

- the dielectric constants, ϵ_{r1} and ϵ_{r2}

- the effective dielectric constants of the single layer cases, $\epsilon_{reff}(h_r = 0)$ and $\epsilon_{reff}(h_r = 1)$

We seek a formula that relates the above parameters of the substates to the transmission line parameters of this structure.

Fig. 2.30 through Fig. 2.32 show the effective dielectric constant as a function of the the height ratio for different dielectric constants of the lower substate in the quasi-static region. Three different normalization schemes are used in the graphs. In Fig. 2.30, a linear normalization is used, described by

$$\epsilon_{reff(linear)}(h_r) = \epsilon_{reff}(h_r) - \epsilon_{reff}(0)$$

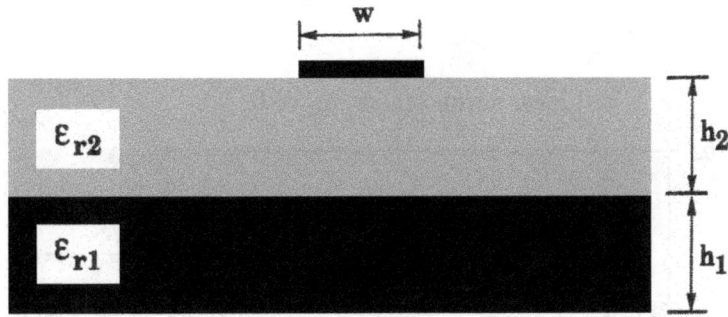

Fig. 2.29: Geometry of coupled microstrips with two substrate layers.

$$- h_r \left[\epsilon_{reff}(1) - \epsilon_{reff}(0) \right] \qquad (2.8)$$

where $\epsilon_{reff}(0)$ and $\epsilon_{reff}(1)$ are the effecitive dielectric constants for the single substrate cases, given by

$$\epsilon_{reff}(0) = \epsilon_{reff}|_{h_1/h_{tot}=0.0} \qquad (2.9)$$

$$\epsilon_{reff}(1) = \epsilon_{reff}|_{h_1/h_{tot}=1.0} \qquad (2.10)$$

These values can be determined from any of the many approximate formulas for the effective dielectric constant of a single substrate structure. While this normalization gives reasonable results when $\epsilon_{r1} > \epsilon_{r2}$, when $\epsilon_{r1} < \epsilon_{r2}$ the results are functions that are much more complex. In addition, this normalization does not result in a fit that is monotonic as a function of the substrate dielectric constant.

In Fig. 2.31, the normalization is based on the capacitance of a parallel plate waveguide with two dielectric layers and is given by

$$\epsilon_{reff(ppwg)}(h_r) =$$

$$\frac{\epsilon_{reff}(0)\epsilon_{reff}(1)}{h_r \left[\epsilon_{reff}(0) - \epsilon_{reff}(1) \right] + \epsilon_{reff}(1)} - \epsilon_{reff}(h_r) \qquad (2.11)$$

Using this normalization gives a better initial approximation to the effective dielectric constant and makes the graphs more uniform for the case when $\epsilon_{r1} < \epsilon_{r2}$. In addition, the normalized functions are monotonic as a function of the substrate dielectric constants, simplifying the curve-fitting procedure.

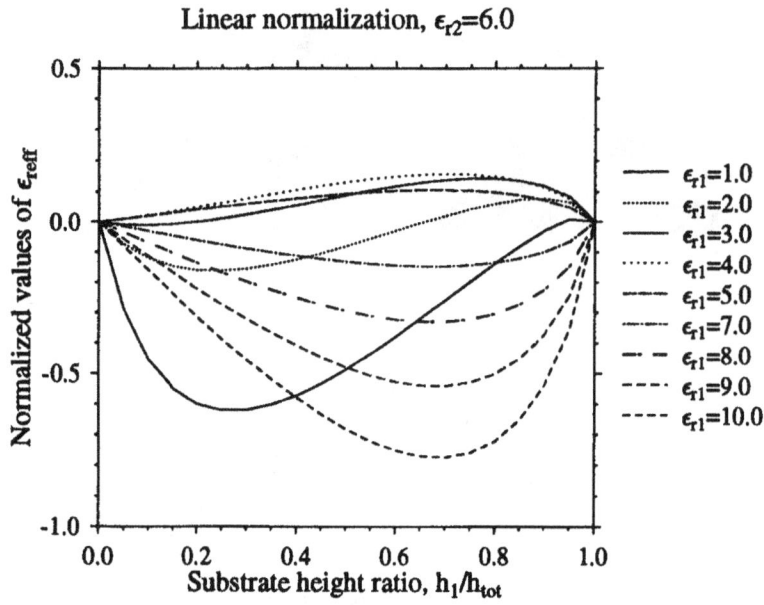

Fig. 2.30: $\epsilon_{r\mathrm{eff}}$ of a microstrip on a two-layer substrate normalized using (2.8) as a function of the substrate height ratio for different values of ϵ_{r1} $(w/h_{tot} = 1.0)$.

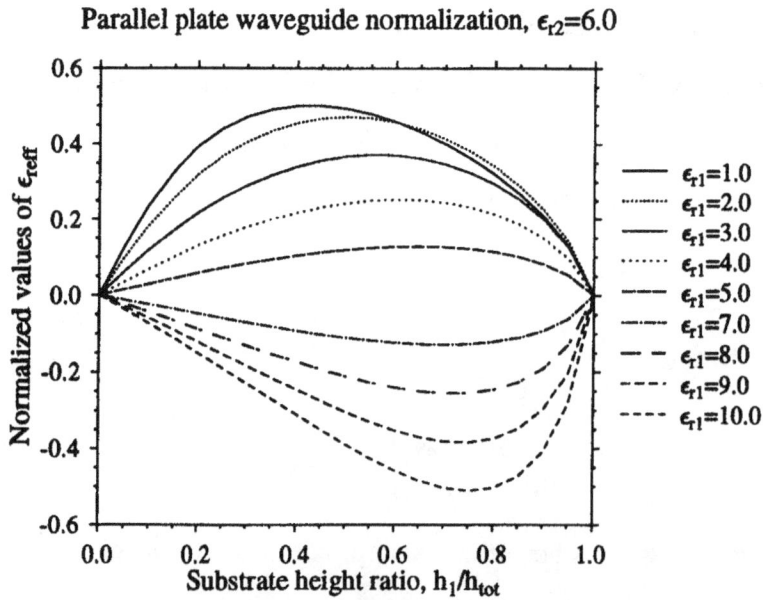

Fig. 2.31: $\epsilon_{r\mathrm{eff}}$ of microstrip on a two-layer substrate normalized using (2.11) as a function of the substrate height ratio for different values of ϵ_{r1} $(w/h_{tot} = 1.0)$.

The final normalization, shown in Fig. 2.32 involves mapping the height ratio such that preceeding equation becomes a linear one. The equation for remapping the height ratio, and the corresponding normalization for ϵ_{reff} are given by

$$h_r' = \left\{ \frac{\epsilon_{reff}(0)\epsilon_{reff}(1)}{h_r\left[\epsilon_{reff}(0) - \epsilon_{reff}(1)\right] + \epsilon_{reff}(1)} - \epsilon_{reff}(0) \right\}$$

$$\cdot \frac{1}{\epsilon_{reff}(1) - \epsilon_{reff}(0)} \tag{2.12}$$

$$\epsilon_{reff(ppwg)}(h_r) = \epsilon_{reff}(h_r) + \epsilon_{reff}(0)$$

$$- h_r'\left[\epsilon_{reff}(1) - \epsilon_{reff}(0)\right] \tag{2.13}$$

This mapping is the one most suited to curve fitting the data, giving an overall relative error that is lower than the error obtained with other techniques using the same order of fitting function.

The full formula for a single microstrip is given by

$$\epsilon_{reff}(h_r) = \frac{\epsilon_{reff}(0)\epsilon_{reff}(1)}{h_r\left[\epsilon_{reff}(0) - \epsilon_{reff}(1)\right] + \epsilon_{reff}(1)}$$

$$+ \left\{ \sum_{n=1}^{5} b_n T_n(h_r')^n + \sum_{n=1}^{2} a_n(\epsilon_{rn} - 1) \right\}$$

$$\cdot h_r'(1 - h_r') \tag{2.14}$$

$$= h_r'\left[\epsilon_{reff}(1) - \epsilon_{reff}(0)\right] + \epsilon_{reff}(0)$$

$$+ \left\{ \sum_{n=1}^{5} b_n T_n(h_r')^n + \sum_{n=1}^{2} a_n(\epsilon_{rn} - 1) \right\}$$

$$\cdot h_r'(1 - h_r') \tag{2.15}$$

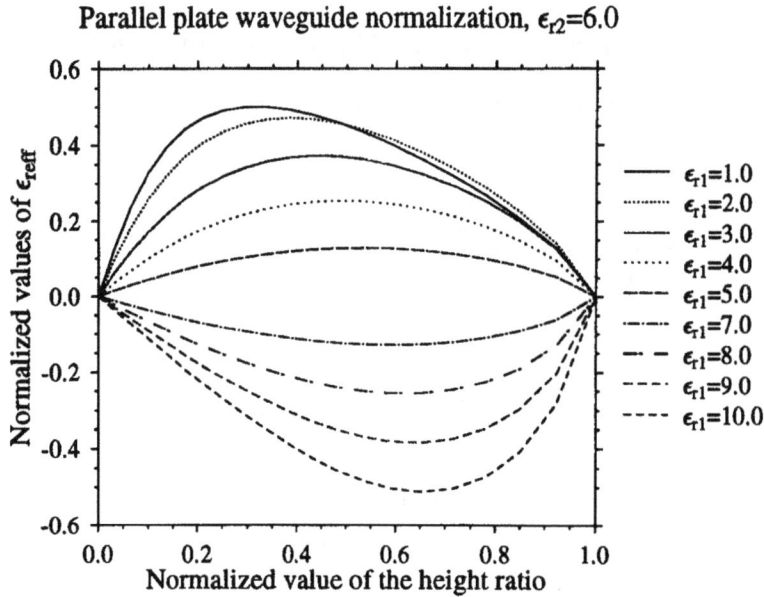

Fig. 2.32: ϵ_{reff} of microstrip on a two-layer substrate normalized using (2.13) as a function of the substrate height ratio normalized using (2.12) for different values of ϵ_{r1} ($w/h_{tot} = 1.0$).

where $T_n(h_r')$ is the Chebyshev polynomial of the first kind of order n. The coefficients in the above expression, b_n and a_n, can be determined using the following relations;

$$b_n = [\epsilon_{reff}(1) - \epsilon_{reff}(0)] \sum_{i=1}^{4} \sum_{j=1}^{5-i} c_{nij} \epsilon_{reff}^i(1) \epsilon_{reff}^j(0) \qquad (2.16)$$

$$a_n = \sum_{i=1}^{4} d_{ni}(w/h_{tot} - 1.0)^i \qquad (2.17)$$

where the constants c_{nij} and d_{ni} are given in Table 1 and Table 2. Most of the dependence on w/h_{tot} is implicitly contained in $\epsilon_{reff}(0)$ and $\epsilon_{reff}(1)$ and so only a small, and simple correction factor is used to improve the accuracy of the solution over a wider range of width to height ratios. The relative error of the function is defined here as

$$(\text{error})_{relative} = 1 - \frac{\epsilon_{reff}^*(h_r, w/h_{tot}, \epsilon_{r1}, \epsilon_{r2})}{\epsilon_{reff}(h_r, w/h_{tot}, \epsilon_{r1}, \epsilon_{r2})} \qquad (2.18)$$

where ϵ_{reff}^* is the approximate value calculated by (2.15) and ϵ_{reff} is the value calculated by the SDA. For $0.2 \leq w/h_{tot} \leq 3.5$, (2.15) gives results that are accu-

Table 2.1: Coeficients, d_{ni}, for equation (2.16)

i	n=1	n=2
1	1.94409×10^{-2}	-5.00773×10^{-3}
2	-7.09419×10^{-3}	4.91136×10^{-3}
3	2.41479×10^{-3}	-2.12875×10^{-3}
4	-3.92206×10^{-4}	3.55627×10^{-3}

rate to within 3 percent relative error over all values of the height ratio and for $1 \le \epsilon_{r1}, \epsilon_{r2} \le 10$.

C. Conclusions

A closed-form expression for the effective dielectric constant of an open microstrip on a two-layer substrate was presented that gives results that have less than three percent relative error. The formula uses the value of the effective dielectric constant of two single layer cases in addition to the physical parameters to provide the accurate results. Results have been validated in the quasi-static region with $0.2 \le w/h_{tot} \le 3.5$, $1 \le \epsilon_{r1}, \epsilon_{r2} \le 10$ and for all values of the height ratio.

Table 2.2: Coeficients, c_{nij}, for equation (2.16)

c_{1ij}			
j	i=1	i=2	i=3	i=4
1	-1.043209	-0.0503390	-0.0797488	0.00798507
2	-0.675972	0.123333	-0.00725776	
3	0.115952	-0.00501308		
4	-0.00759135			

c_{2ij}			
j	i=1	i=2	i=3	i=4
1	-0.146779	0.474085	-0.0278816	0.00152566
2	0.539591	-0.142693	0.00435987	
3	-0.103164	0.0104707		
4	0.00659505			

c_{3ij}			
j	i=1	i=2	i=3	i=4
1	0.427940	-0.976178	0.0920436	-0.00530197
2	-0.538552	0.235510	-0.00943642	
3	0.0863343	-0.0152405		
4	-0.00494465			

c_{4ij}			
j	i=1	i=2	i=3	i=4
1	-0.04520958	0.388317	-0.0176737	7.35419×10^{-4}
2	0.109800	-0.107298	0.00282642	
3	-0.00953375	0.00784613		
4	-4.78285×10^{-5}			

c_{5ij}			
j	i=1	i=2	i=3	i=4
1	0.0749343	-0.233050	0.0164757	-8.25541×10^{-4}
2	-0.0811032	0.0594164	-0.00200066	
3	0.00974942	-0.00406101		
4	-3.52724×10^{-4}			

CHAPTER 3

ANALYSIS OF 3-D ELECTRONIC PACKAGING CIRCUITS

3.1 Introduction

Recent advancements in Monolithic Microwave Integrated Circuit (MMIC) technology resulted in electronic packages of significantly smaller size and a larger number of printed interconnects on the motherboard. Accurate design and optimization of such high-speed high-density microwave circuits becomes a major challenge when it comes to high performance and low cost. High frequency operation is usually the main cause of strong coupling and interference among neighboring transmission lines, thereby affecting the overall electrical performance of the package. The presence of abrupt discontinuities, microstrip bends, bond wires, metallic bridges and vertical conducting vias results in additional parasitic effects such as radiation and time delays. The major objective of current technology is the design of electronic packages that are optimized to minimize severe parasitic effects without necessarily increasing the cost or the complexity of the manufacturing process.

The existing high demand for the development of more accurate, versatile and efficient numerical models which can be used in the design and characterization of microwave circuits mandates the implementation of full-wave techniques such as the Finite-Difference Time-Domain (FDTD) method [47], the Spectral Domain Approach (SDA) [48], and the Finite Element Method (FEM) [49]. The FDTD method is probably the most extensively used technique for the analysis of geometrically complex packaging structures. It was initially applied for the evaluation of frequency dependent parameters of basic microstrip discontinuities [50],[51]. It was later implemented successfully in the analysis of more complex structures such as filters, microstrip transitions, bond wires, bridges, *etc.* [52],[53]. However, the main drawback of the method is that curved surfaces and non-rectangular volumes are usually modeled using a staircasing approach. The SDA technique is also very popular in the area of microwave circuit analysis and design. Its main disadvantage though is that it can only handle metalizations in the same plane. Although the method can be extended to treat conducting transitions in the vertical plane [54], it still is restricted to specific type of geometries. On the contrary, the FEM is the most versatile and flexible numerical technique to be used in the analysis of geometrically complicated electronic packaging structures. The introduction of vector finite elements [55], the valuable contributions on Absorbing Boundary Conditions (ABC) [56] and the effectiveness of sparse matrix iterative solvers, created a conducive environment for the evolution of the method in the area of computational electromagnetics. The FEM has been extensively used for scattering and radiation problems [57], waveguide propagation problems [58], and analysis of two-dimensional (2-D) MMIC structures [12],[59]. Recently, the method has been applied in the S-parameter evaluation of three-dimensional (3-D) MMICs such as microstrip transitions, planar discontinuities and conducting vias [60]-[62].

This study formulates a full-wave analysis and implementation of the finite element method to solve geometrically complex and practical microwave circuits. Unlike previous work done on the subject [60]-[62], a 2-D eigenvalue analysis [59] is now performed at the input port to compute the field distribution of the dominant or higher order modes; the circuit discontinuity is then excited with the governing modal distribution. The eigenvalue analysis is also applied to the output port in order to calculate the frequency dependent propagation constant and characteristic impedance of the transmission line. The dispersive propagation constant at the input and output ports is used in the implementation of the ABCs, whereas the characteristic impedance is used in the evaluation of the S-parameters. The current formulation is proven to be *efficient*, *flexible* and extremely *accurate* in analyzing complex 3-D microwave circuits. It is *efficient* because of the use of a 2-D eigenvalue analysis to determine the excitation fields and needed circuit parameters. It is *versatile* because the input and output ports are not restricted to a single microstrip line; coplanar waveguides, coupled microstrip lines and finlines can also be used. It is *accurate* because the excitation field, propagation constant and characteristic impedance are computed at every frequency using a full-wave approach.

3.2 The Finite Element Method

A full-wave finite element method is used in the analysis of complex electronic packaging circuits printed on single or multi-layer substrates. A typical microstrip discontinuity is illustrated in Fig. 3.1. The input port of the structure is excited using the dominant field distribution at a specific frequency. The governing field distribution at the input port is evaluated apriori using a two-dimensional eigenvalue analysis [59]. Both input and output ports are appropriately terminated using absorbing boundary conditions that are directly applied to the transverse electric field component at the surface. The same type of absorbing boundary conditions are also used to effectively terminate the side walls of open structures.

The three-dimensional finite element analysis starts with the discretization of Helmholtz's equation in a source free region

$$\nabla \times \left([\mu_r]^{-1} \cdot \nabla \times \mathbf{E}\right) - k_o^2 \left[\epsilon_r\right] \mathbf{E} = 0, \tag{3.1}$$

where $[\epsilon_r]$ and $[\mu_r]$ are, respectively, the relative permittivity and permeability tensors of the domain. Using the well-known Galerkin's technique, (3.1) may be transformed in a weak integral form given by

$$\int_{\Omega} ([\mu_r]^{-1} \nabla \times \mathbf{E}) \cdot (\nabla \times \mathbf{N}) \, dV - k_o^2 \int_{\Omega} [\epsilon_r] \cdot \mathbf{E} \cdot \mathbf{N} \, dV + \oint_S ([\mu_r]^{-1} \nabla \times \mathbf{E}) \cdot (\mathbf{N} \times \hat{a}_n) \, dA = 0$$
$$\tag{3.2}$$

where $\hat{a}_n$ is the normal to the surface unit vector pointing outside the finite element volume, and $\mathbf{N}$ denotes a given vector testing function. The closed surface integral in (3.2) is nonzero only on nonperfectly conducting surfaces.

By discretizing the integral equation in (3.2) with the use of tetrahedral elements, the electric field $\mathbf{E}$ is expanded in terms of a set of vector basis functions $\mathbf{N}$ to finally

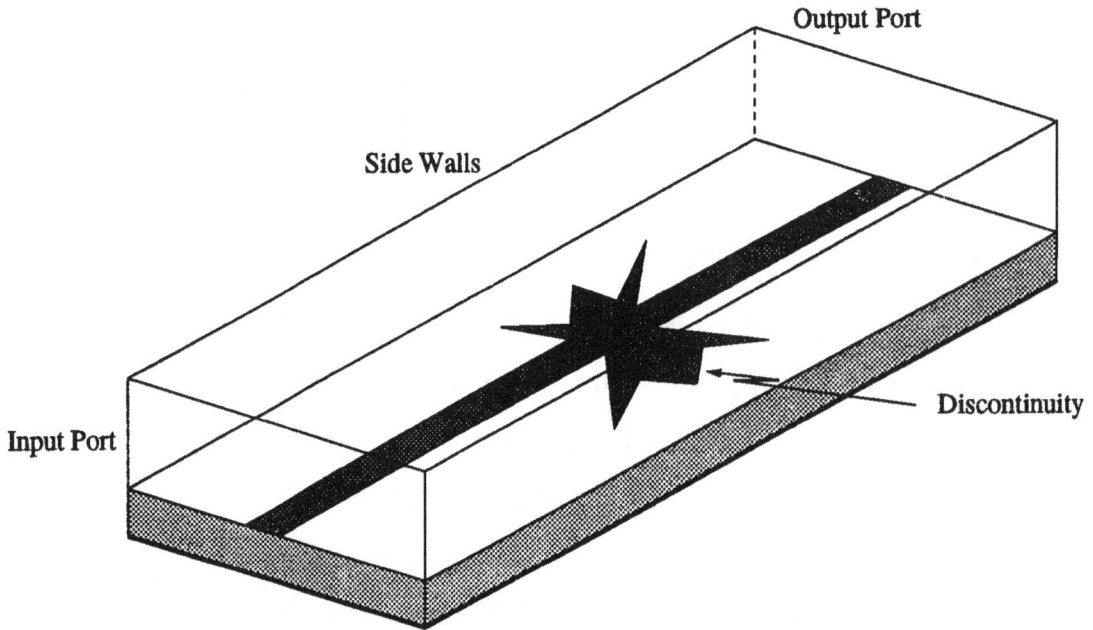

Fig. 3.1: Three-dimensional rendering of a typical microstrip discontinuity.

obtain the following elemental matrix system:

$$[M^e + B^e_{S_1} + B^e_{S_2} + B^e_{S_3}]\{E^e\} = \{b^e\} \tag{3.3}$$

where

$$M^e(i,j) = \int_{\Omega^e} \{\nabla \times \mathbf{N}_i\}^T \cdot [\mu_r]^{-1} \cdot \{\nabla \times \mathbf{N}_j\} \, dV - k_o^2 \int_{\Omega^e} \mathbf{N}_i^T \cdot [\epsilon_r] \cdot \mathbf{N}_j \, dV \tag{3.4}$$

$$B^e_{S_1}(i,j) = jk_z \int_{S_1^e} \{\mathbf{N}_i \times \hat{a}_{n_1}\}^T \cdot [\mu_r]^{-1} \cdot \{\mathbf{N}_j \times \hat{a}_{n_1}\} \, dA \tag{3.5}$$

$$B^e_{S_2}(i,j) = jk_n \int_{S_2^e} \{\mathbf{N}_i \times \hat{a}_{n_2}\}^T \cdot [\mu_r]^{-1} \cdot \{\mathbf{N}_j \times \hat{a}_{n_2}\} \, dA \tag{3.6}$$

$$B^e_{S_3}(i,j) = jk_o\sqrt{\epsilon_r\mu_r} \int_{S_3^e} \{\mathbf{N}_i \times \hat{a}_{n_3}\}^T \cdot [\mu_r]^{-1} \cdot \{\mathbf{N}_j \times \hat{a}_{n_3}\} \, dA \tag{3.7}$$

$$b^e(i) = \int_{S_1^e} \{\mathbf{N}_i \times \hat{a}_{n_1}\}^T \cdot [\mu_r]^{-1} \cdot \{\mathbf{E}_t^{inc} \times \hat{a}_{n_1}\} \, dA \tag{3.8}$$

where $i,j = 1, ..., 6$. The elemental matrices are then assembled into the global matrix using the edge connectivity information. The global matrix system is solved using an efficient Conjugate Gradient Square (CGS) solver with Jacobi preconditioning. For better and faster convergence, the iteration is done in double precision.

Once the electric field distribution is obtained everywhere in the structure, the next step is to evaluate the corresponding voltages at the two ports. Note that although the theoretical formulation was based on a two-port network, it can be easily extended for multiple ports. The S-parameters of the structure are evaluated using

$$S_{11} = \frac{V_1 - V_1^{ref}}{V_1^{ref}} \qquad (3.9)$$

$$S_{21} = \frac{V_2}{V_1^{ref}} \sqrt{\frac{Z_{c1}}{Z_{c2}}} \qquad (3.10)$$

where V_1 and V_2 are the voltages calculated at ports 1 and 2 (3-D analysis), respectively, whereas V_1^{ref} is the reference voltage calculated at port 1 (2-D analysis). Also, Z_{c1} and Z_{c2} are the corresponding characteristic impedances of the transmission lines at the two ports. These were calculated apriori using the 2-D finite element eigenvalue analysis.

3.3 The Finite-Difference Time-Domain Method

The Finite-Difference Time-Domain (FDTD) method is extensively used in this study to verify some of the predicted data obtained using the FEM. In this section, the method is briefly introduced by presenting some of the most important details in terms of its implementation in packaging and microwave circuits.

The FDTD method is one of the most popular numerical techniques for solving complex electromagnetic problems. The FDTD method is finding applications in a wide spectrum of simulation problems including antennas for wireless communications, biomedical applications, microwave circuits, electronic packaging and electromagnetic scattering and penetration. The popularity of this method is attributed to its simplicity in implementation and computer programming, its ability to handle arbitrary and complex geometries including different materials, and the fact that it is a time-domain method. Frequency information is obtained through a single simulation over a broad frequency spectrum.

An FDTD code suitable for analyzing general multi-conductor structures has been developed. The code is quite general in terms of modeling different material and conductor discontinuities, such as the ones found in electronic packages. The developed code uses first-order Mur absorbing boundary conditions. These have been proven to work well in applications involving microwave circuits. The electric wall source condition has been implemented to excite the dominant mode of structures investigated in this paper. Since the FDTD method uses rectangular bricks as the basic mesh elements, it is predominantly suited for planar structures. For structures characterized by curved surfaces the FEM is more suitable.

In obtaining the S-parameters of a given structure, a source plane is imposed at the input port. The excitation signal is a Gaussian pulse in the time domain. Once the pulse is launched, the first-order Mur absorbing boundary conditions are immediately turned on. The numerical simulation is carried out twice. The first simulation occurs in the absence of the discontinuity. This is required in order to establish a reference incident waveform propagating along the microstrip line. The reference plane is defined N cells away from the beginning of the discontinuity. A second simulation is repeated in the presence of the discontinuity and the time signature of the incident and reflected voltages at the reference plane is obtained. Using the two simulations, the incident and reflected time-domain waveforms are first calculated, and then used to evaluate the amplitude and phase of the return loss. A similar argument holds for the transmitted voltage used in the evaluation of the insertion loss of the structure.

3.4 Numerical Validation and Results

The finite element formulation presented in Section 3.2 was successfully implemented and applied to a variety of three-dimensional circuits that are frequently used in microwave electronic packages. The finite element method was extensively verified by comparing with results obtained using the finite-difference time-domain method briefly outlined in Section 3.3.

The first geometry considered, primarily for verification purposes, is the rectangular microstrip patch antenna shown in Fig. 3.2. The patch antenna is printed on a RT/Duroid substrate with $\epsilon_r = 2.2$ and height 0.794 mm. A 50 Ohms microstrip line is used to excite the patch. The same exact geometry was analyzed in the past by Sheen *et al.* [52] using the finite-difference time-domain method. The same mesh sizes suggested in [52] were also used here; *i.e.*, $\Delta x = 0.389$ mm, $\Delta y = 0.4$ mm, $\Delta z = 0.265$ mm and $\Delta t = 0.441$ ps. These mesh sizes result in an integral number of cells along the width and length of the patch, but not along the width of the microstrip line feeding the patch. The resulting FDTD mesh dimensions are $61 \times 100 \times 17$ cells.

The Return Loss (RL) obtained using the FEM and the FDTD method is shown in Fig. 3.3. A fairly good agreement between the two methods is illustrated. For frequencies lower than 10 GHz, where the mesh density is sufficiently fine, the agreement between the two numerical techniques is excellent. Two different finite element discretizations were considered: one with $22,702$ tetrahedras and the other with $28,883$ tetrahedras. However, as shown in Fig. 3.3, only a minor improvement is observed in the predictions when running the largest discretization. A possible source of error in the calculations is the inability of the FDTD method to properly match all microstrip surface dimensions. A narrower microstrip line for example, always results in a larger characteristic impedance, thereby affecting the return loss of the structure, especially at the higher frequencies. On the other hand, using the FEM, all geometry dimensions are precisely modeled.

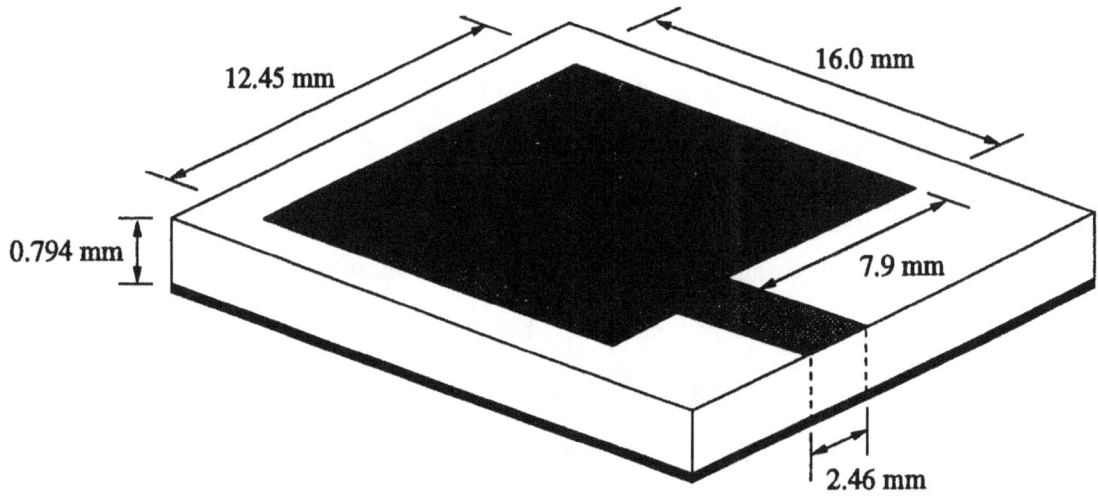

Fig. 3.2: Geometry of a rectangular microstrip patch antenna on a RT/Duroid substrate with $\epsilon_r = 2.2$.

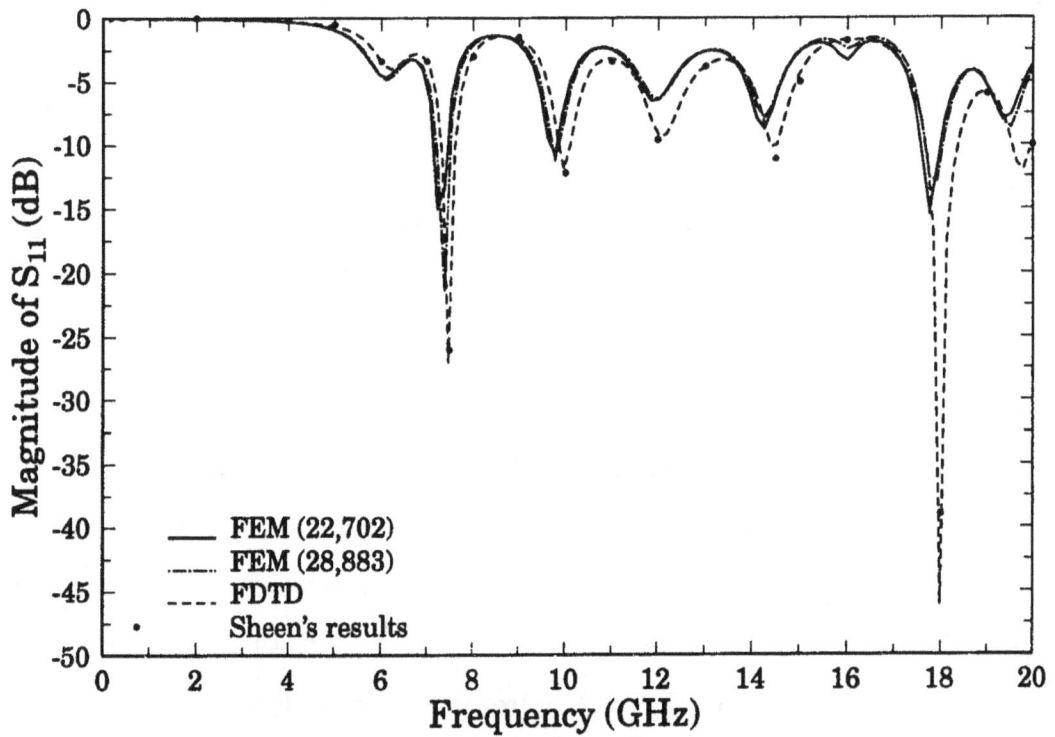

Fig. 3.3: Return loss of a rectangular microstrip patch antenna.

In order to provide insight into the computational effort required by the FEM, the following statistics were reported. The original mesh consisted of $22,702$ tetrahedras and a total of $25,625$ unknowns. The computational time was approximately 30 minutes per frequency point in the lower frequency range and 15 minutes per frequency point in the upper frequency range. This problem was run on a 370 IBM RISC/6000 UNIX workstation. The solution tolerance based on the residual norm was set to $1.0e-6$. The recorded computational time also accounts for the CPU time needed in evaluating the modal field distribution at the input port. On the other hand, the FDTD code took approximately 45 minutes for the overall simulation; $8,192$ time steps were allowed for the pulse to propagate out. The simulation was done on a Silicon Graphics Power Indigo2 workstation with an R8000 processor. Note that the latter is a significantly faster computer than the 370 IBM RISC/6000.

The second circuit considered in this study was also extracted from the paper by Sheen *et al.* [52]. This is the low-pass filter illustrated in Fig. 3.4. The perfectly conducting microstrip surfaces are printed on a RT/Duroid substrate with $\epsilon_r = 2.2$ and height 0.794 mm. This geometry was run using both the FEM and the FDTD codes for a frequency range of 20 GHz. The magnitude of S_{11} and S_{21} versus frequency is illustrated in Figs. 3.5 and 3.6, respectively, whereas the corresponding phases are illustrated in Figs. 3.7 and 3.8. The phase of S_{11} was evaluated at a distance 4.233 mm away from the discontinuity; the phase of S_{21}, on the other hand, was evaluated 3.3864 mm away from the discontinuity. All four figures show an excellent agreement between the two methods. The finite element mesh consisted of $28,914$ tetrahedras and a total of $33,532$ unknown field components. The corresponding CPU time for this problem was approximately 20 minutes per frequency point in the lower frequency range and 10 minutes per frequency point in the upper frequency range. Note that although the low-pass filter is computationally a larger problem than the microstrip patch antenna, the required CPU time is significantly less. The reason is related to the condition number of the resulting matrix system. As far as the FDTD method is concerned, the mesh dimensions were the following: $\Delta x = 0.4064$ mm, $\Delta y = 0.4233$ mm, $\Delta z = 0.265$ mm and $\Delta t = 0.441$ ps. The overall mesh size was $81 \times 101 \times 17$ cells. The required computational time was approximately 50 minutes; again, a total of $8,192$ time steps were allowed for the pulse to propagate out. The simulation was run on a Silicon Graphics Power Indigo2 workstation with an R8000 processor.

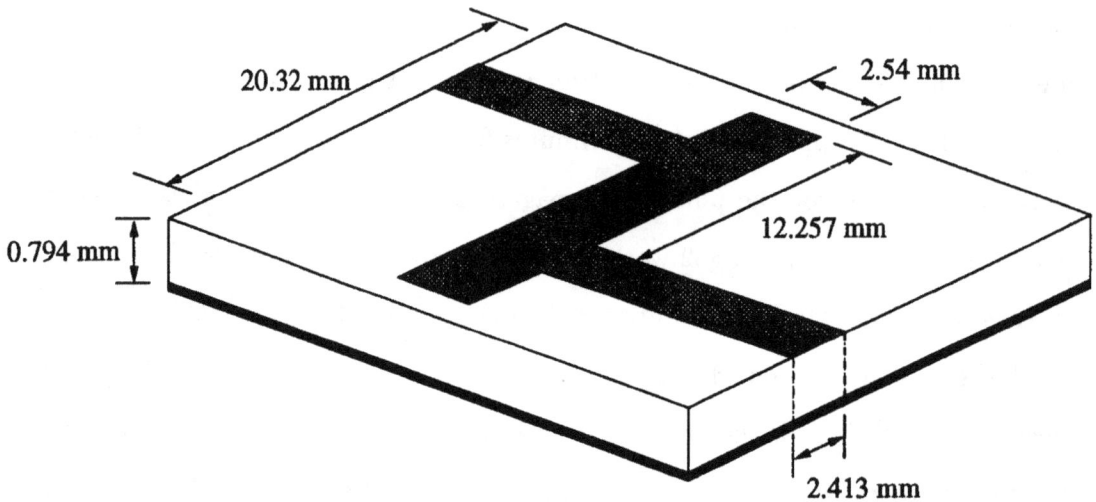

Fig. 3.4: Geometry of a low-pass filter on a RT/Duroid substrate with $\epsilon_r = 2.2$.

Although spiral inductors, as well as inductors in general, are commonly found in microwave circuits, most of the relevant analysis has been done based on either quasi-static methods [63] or rectangular grid methods such as FDTD, SDA, Transmission Line Method (TLM), and Method of Lines (MoL) [64]. Using the FEM, electromagnetic modeling of curved structures becomes only a matter of simply drawing and discretizing the geometry. A circular spiral inductor connected in series with a microstrip line on an Alumina substrate with $\epsilon_r = 9.8$ is shown in Fig. 3.9. One end of the spiral is bonded with the microstrip line at port 2 through a cylindrical metallic bridge. The bridge surface is defined by three points: the first point is in the center of the spiral, the second point is at the edge of the microstrip line, and the third point is in the middle of the gap (height of 1.0 mm). The spiral is made out of $1\frac{1}{2}$

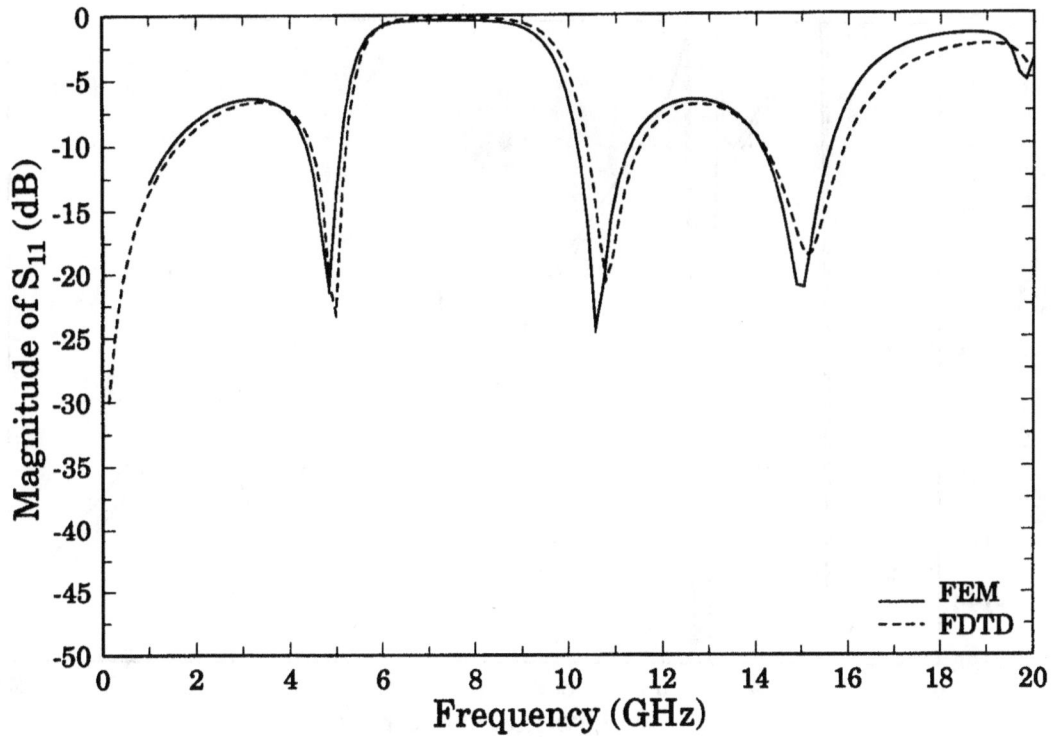

Fig. 3.5: Return loss of a low-pass filter.

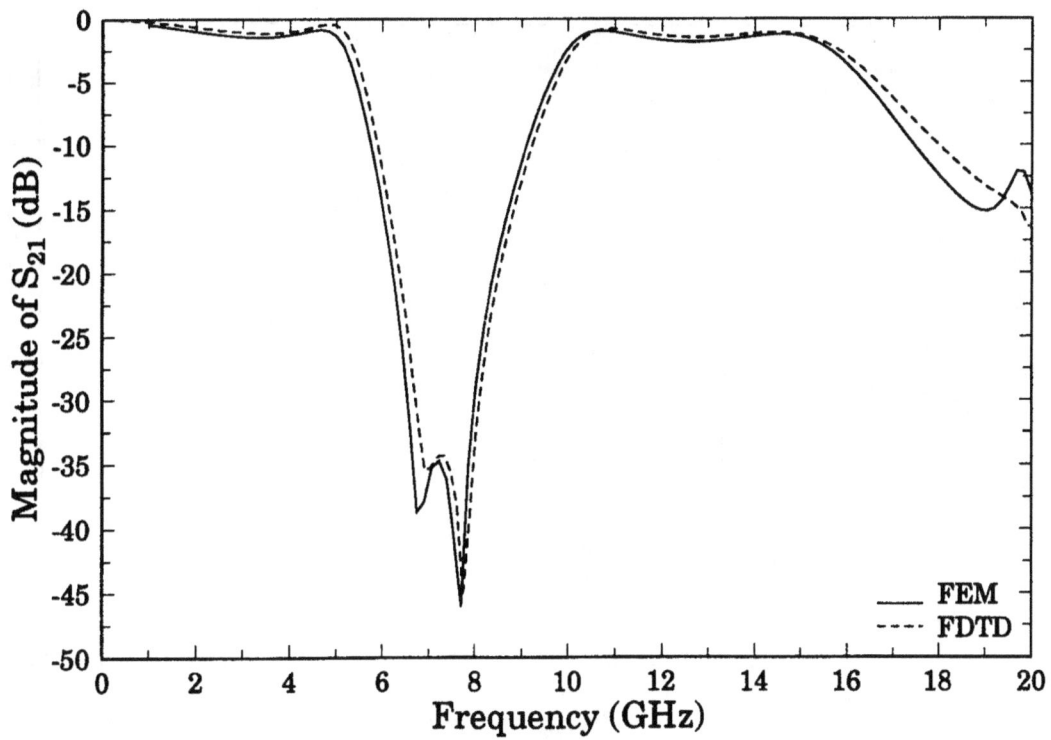

Fig. 3.6: Insertion loss of a low-pass filter.

63

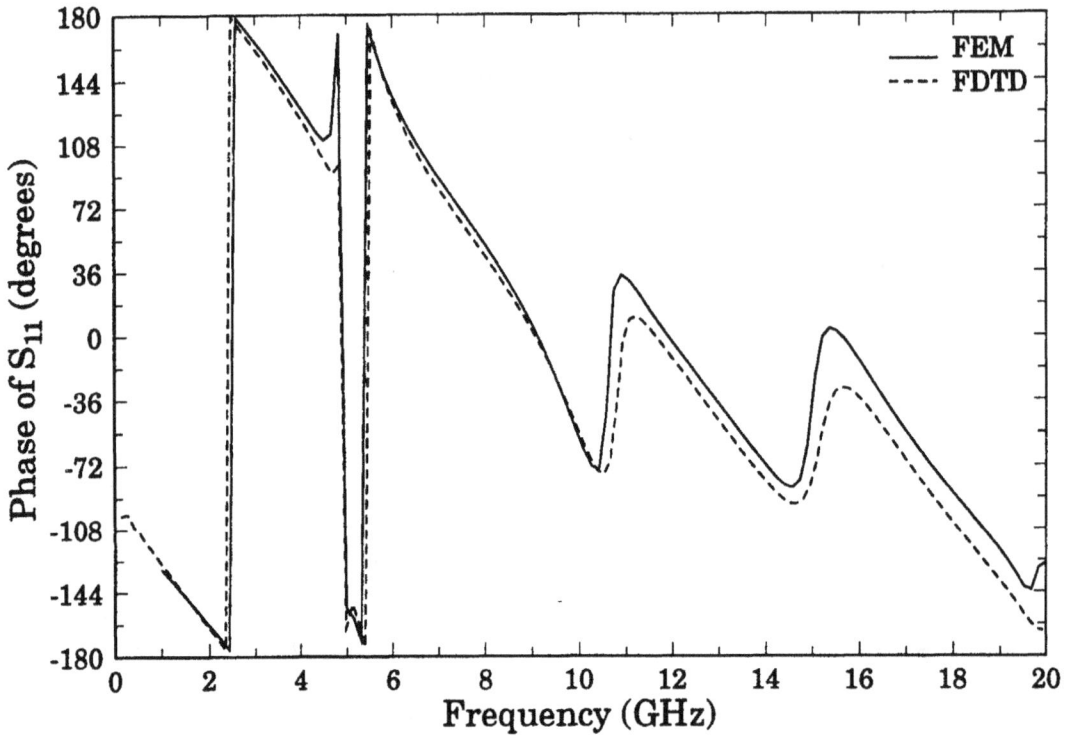

Fig. 3.7: S_{11} phase of a low-pass filter.

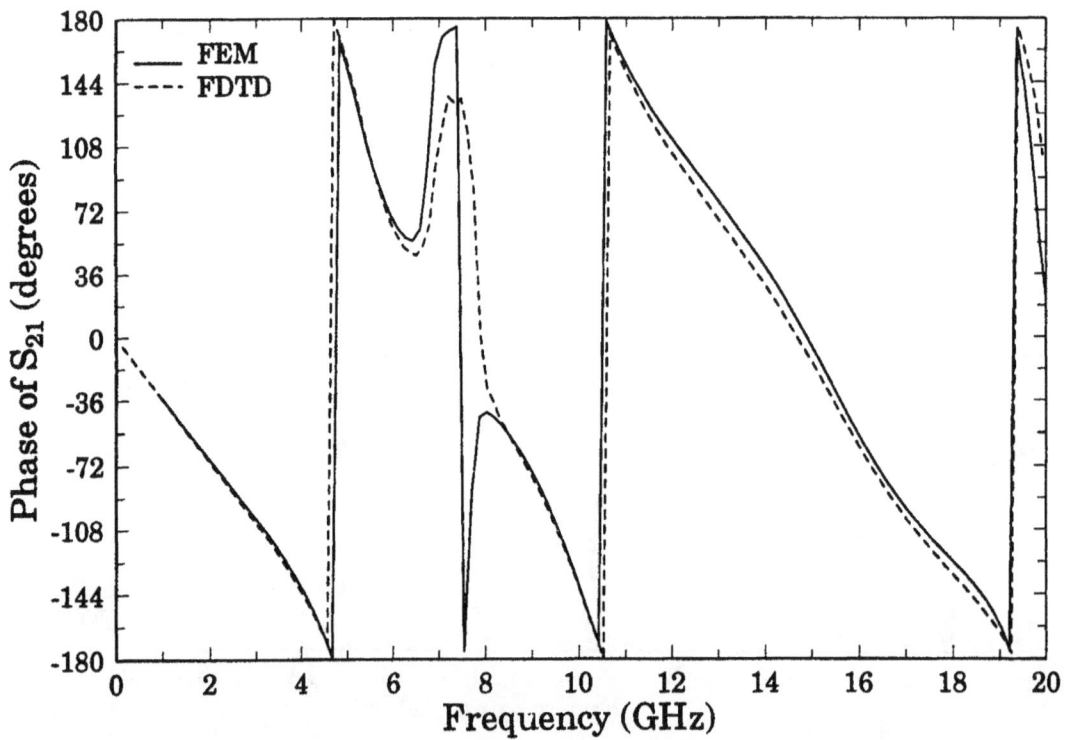

Fig. 3.8: S_{21} phase of a low-pass filter.

turns with width 0.2 mm. The microstrip line at the input and output ports is 0.635 mm wide, and the substrate height is also 0.635 mm. The magnitude of S_{11} and S_{22} calculated using the FEM is illustrated in Fig. 3.10. Although measurements were not available for data comparison, the geometry discretization in both cases was sufficiently fine to ensure accurate simulations. Specifically, the mesh consisted of 26,096 elements and a total of 31,093 unknowns. The corresponding computational time was approximately an hour per frequency point in the lower range of frequencies, and 20 minutes per frequency point in the intermediate-to-upper range of frequencies; again, a 370 IBM RISC/6000 workstation was used to run this problem. Most of the computational effort (90%) was spent solving the linear system of equations. Comparing the two plots in Fig. 3.10, it is interesting to observe that those are not identical, although very similar. The minor differences are attributed basically to the presence of the cylindrical metallic bridge. In addition, it is important to mention here that the spiral inductor behaves as a lumped element only in the lower range of frequencies (linear region); at higher frequencies the structure begins to resonate due to additive capacitive effects.

A spiral inductor is usually connected either in series or in parallel. The same configuration as the one used in the previous example is now connected in shunt with a microstrip line printed on an Alumina substrate. The geometry and dimensions of the structure are shown in Fig. 3.11. The center of the spiral is grounded using a planar conducting via. The magnitude plots of S_{11} and S_{21} calculated using the FEM code are shown in Fig. 3.12. Although comparisons are not available, it is interesting to observe that at lower frequencies the structure indeed behaves as a lumped inductor connected in shunt. Such structure though is highly resonant; therefore, multiple peaks and nulls appear in the higher frequency range. As a result, the resulting S-parameters are plotted only up to 7 GHz. The finite element mesh for this problem consisted of 43,588 tetrahedras and a total of 51,270 unknowns.

Spiral inductors are used in microwave circuits to simply introduce a relatively

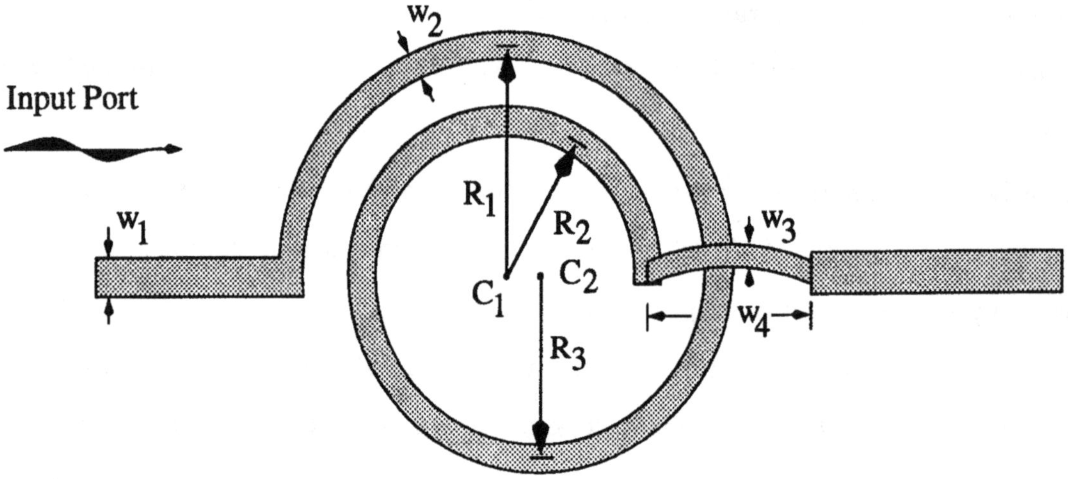

Fig. 3.9: Series spiral inductor printed on an Alumina ($\epsilon_r = 9.8$) substrate of height 0.635 mm. The end of the spiral is bonded with a microstrip line through a metallic bridge of height 1.0 mm. The bridge has an arc shape defined by three points. Other dimensions: $w_1 = 0.635$ mm, $w_2 = w_3 = 0.2$ mm, $w_4 = 2.3$ mm, $R_1 = 1.9$ mm, $R_2 = 1.3$ mm, $R_3 = 0.7$ mm.

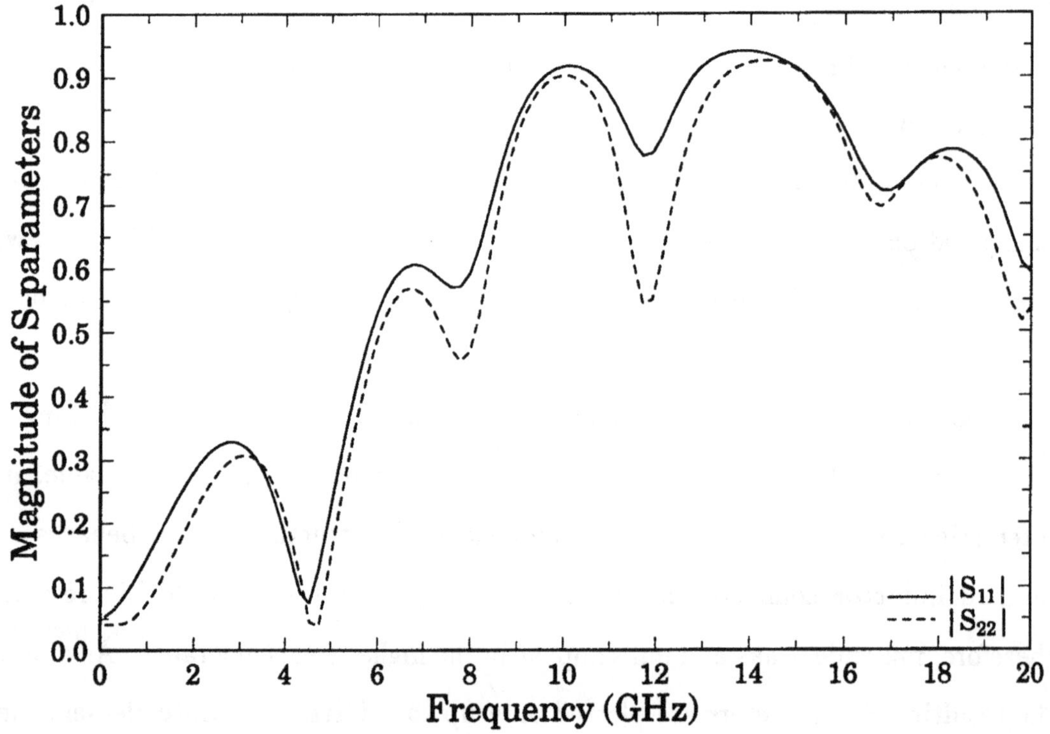

Fig. 3.10: Magnitude of S_{11} and S_{22} for a series spiral inductor with a bond wire bridge.

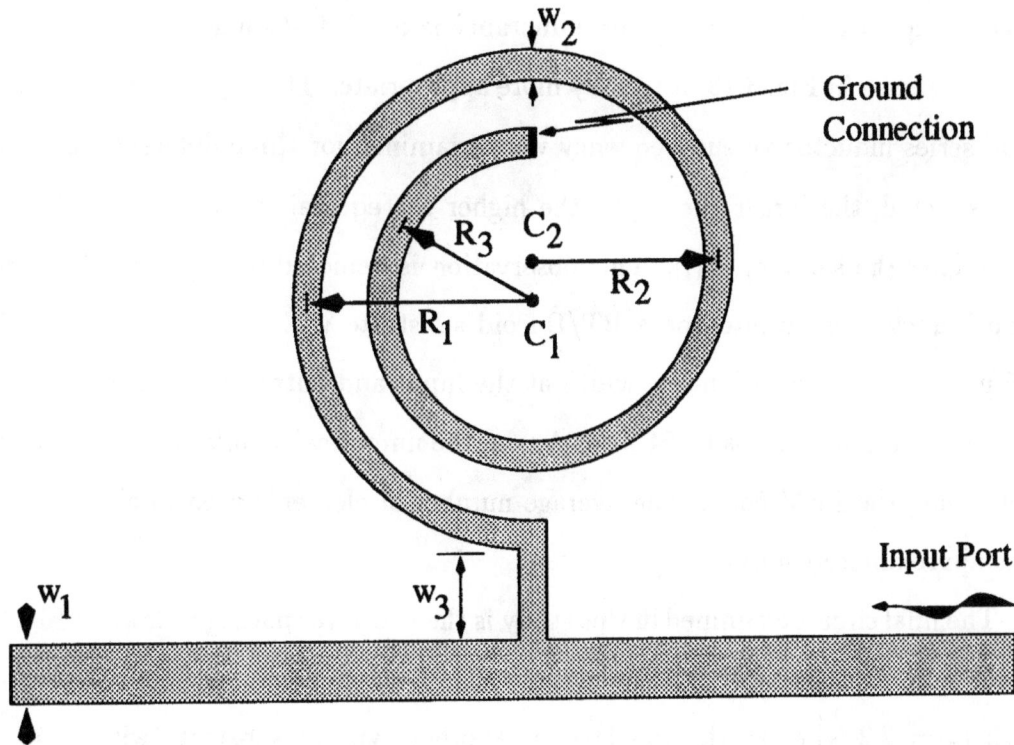

Fig. 3.11: Spiral inductor connected in shunt with a microstrip line printed on an Alumina ($\epsilon_r = 9.8$) substrate of height 0.635 mm. Other dimensions: $w_1 = 0.635$ mm, $w_2 = 0.2$ mm, $w_3 = 0.6$ mm, $R_1 = 1.9$ mm, $R_2 = 1.3$ mm, $R_3 = 0.7$ mm.

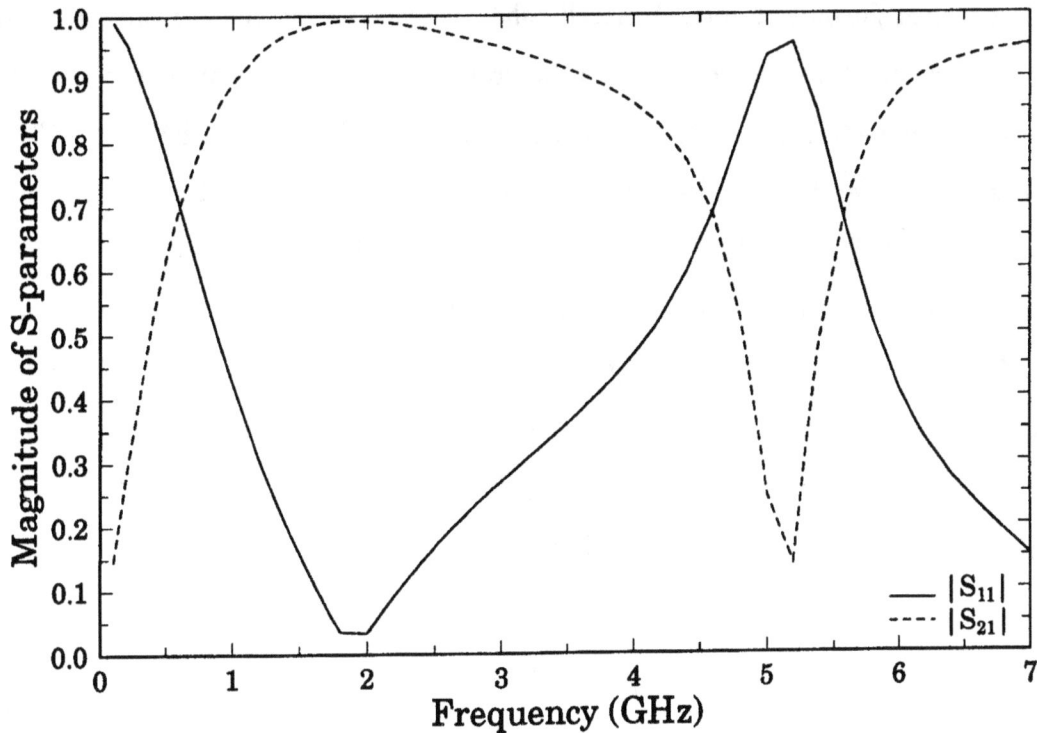

Fig. 3.12: Magnitude of S_{11} and S_{22} for a spiral inductor connected in shunt across a microstrip line.

high inductance. When a smaller inductance is needed, a single loop inductor, like the one shown in Fig. 3.13, is usually more appropriate. The S-parameters of a single loop series inductor versus frequency were examined for three different values of ϕ. As expected, the larger the angle, the higher the equivalent inductance; therefore, the higher the slope of $|S_{11}|$. This observation is depicted in Fig. 3.14. The single loop inductor was printed on a RT/Duroid substrate with $\epsilon_r = 2.2$ and line width 0.5 mm. The transmission line width at the input and output ports is 2.4 mm. The height of the substrate is 0.794 mm. Again, the numerical simulation was performed only using the FEM code. The average number of elements used in all three cases was approximately $40,000$.

The final circuit examined in this study is the two-layer package shown in Fig. 3.15 which is representative of practical designs. The bottom layer is a dielectric substrate with $\epsilon_r = 2.2$ whereas the top layer is another type of substrate with $\epsilon_r = 6.2$. The microstrip at the input port is connected to the microstrip at the upper layer through a vertical conducting via; an identical via joins the upper microstrip with the microstrip at the output port. A metallic sheet is placed at the bottom interface of the upper dielectric layer to provide potential grounding. Also, all geometry shapes were chosen to be rectangular so the simulation is performed with both FEM and FDTD codes. The magnitude of S_{11} versus frequency, calculated using the two numerical techniques, is illustrated in Fig. 3.16. Although both methods accurately predict the resonant frequency of the package, there are some minor discrepancies between the two data sets. The reason might be attributed to non-physical reflections from the surrounding absorbing boundary conditions. More efficient mesh terminations, such as the recently developed Perfectly Matched Layer (PML) [65]-[67], can be implemented in both FEM and FDTD methods to further reduce possible truncation error.

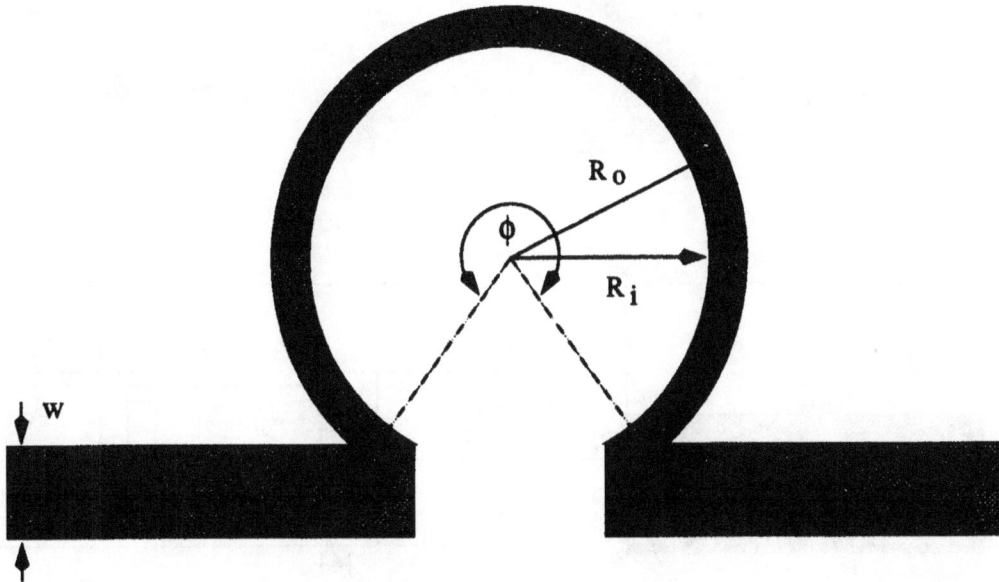

Fig. 3.13: Single-loop series inductor printed on a RT/Duroid substrate with dielectric constant $\epsilon_r = 2.2$ and height $h = 0.794$ mm. Other dimensions: $w = 2.4$ mm, $R_i = 1.5$ mm, $R_o = 2.0$ mm.

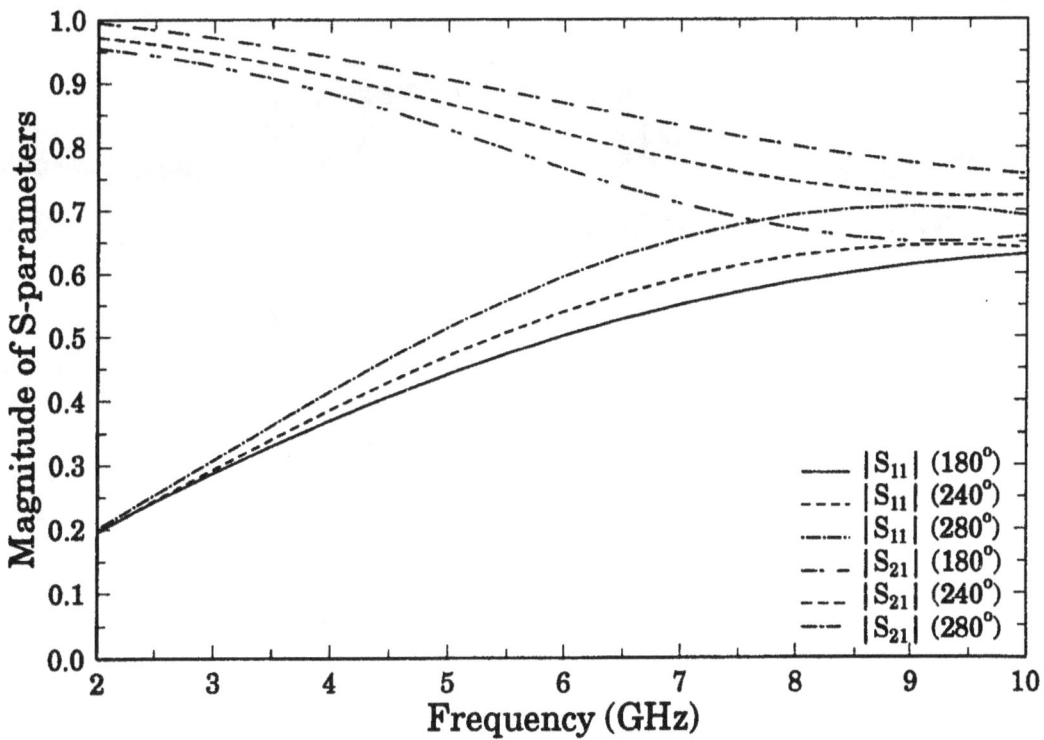

Fig. 3.14: S-parameters of a single-loop series inductor as a function of the angle ϕ.

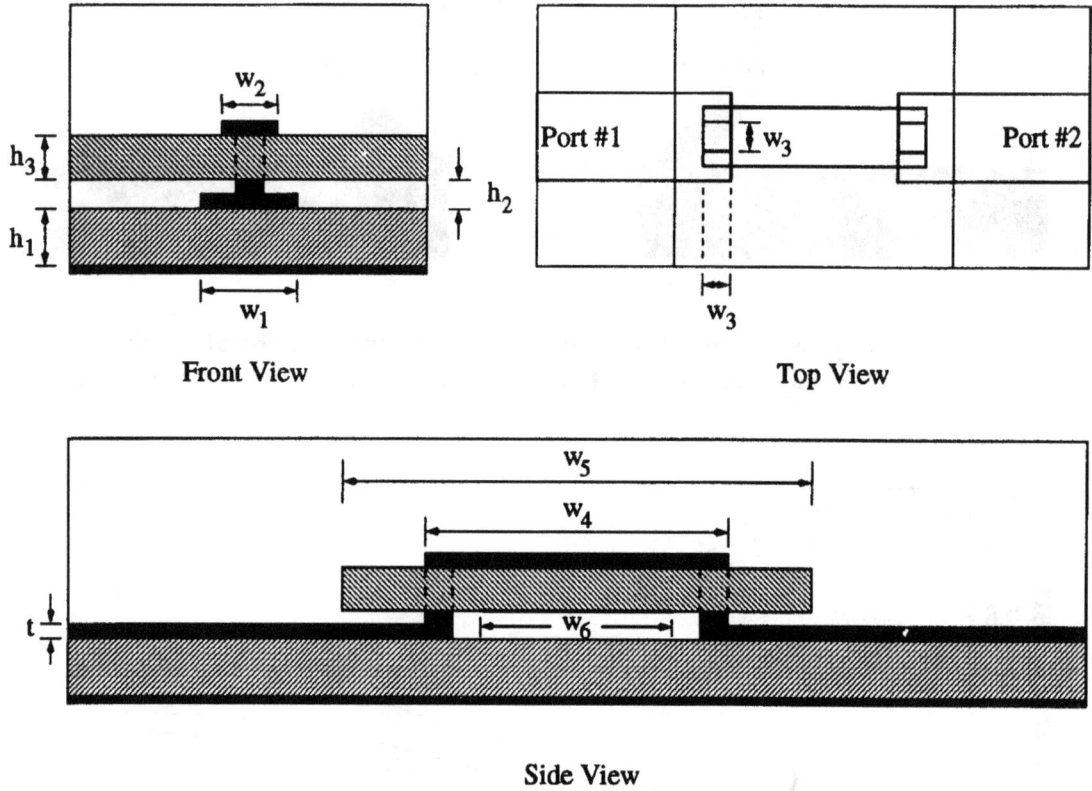

Front View

Top View

Side View

Fig. 3.15: Double via transition between two microstrip lines printed on different substrate layers. The bottom layer is a dielectric material with $\epsilon_r = 2.2$ and the top layer is another type of material with $\epsilon_r = 6.2$. Dimensions: $h_1 = 0.8$ mm, $h_2 = 0.4$ mm, $h_3 = 0.6$ mm, $w_1 = 2.4$ mm, $w_2 = 0.8$ mm, $w_3 = 0.4$ mm, $w_4 = 5.2$ mm, $w_5 = 6.8$ mm, $w_6 = 3.6$ mm, $t = 0.2$ mm.

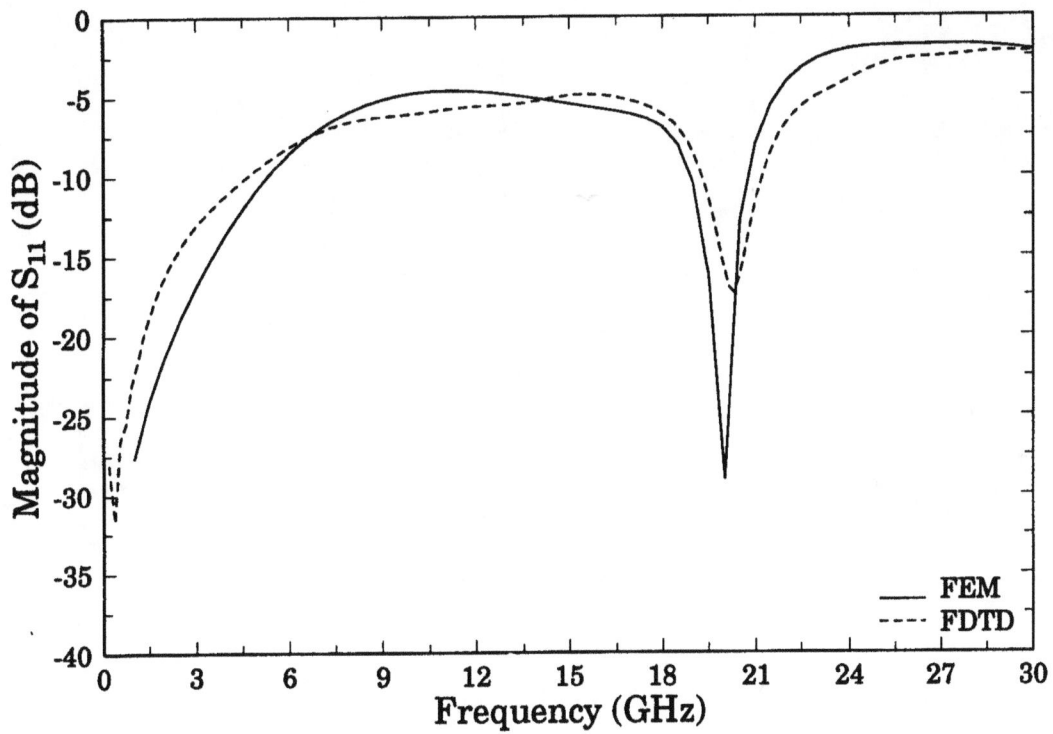

Fig. 3.16: S_{11} versus frequency for the double-via-transition package.

CHAPTER 4

ANALYTICAL ASYMPTOTIC EXTRACTION TECHNIQUES FOR MULTI-LAYER TRANSMISSION LINES AND PLANAR CIRCUITS

4.1 Introduction

Accurate and efficient analysis of planar transmission lines and printed circuits is an essential part of realizing successful designs of high frequency MMIC's circuits. In early integrated circuit designs, the quasi-static method based on low frequency assumptions was used with some success. However, with recent advances in MMIC technologies, the operating frequency and the circuit density continue to increase. Therefore more accurate and efficient models are required to enhance circuit performance. These requirements necessitate full-wave analyses which take into account radiation, surface waves, and mutual coupling between lines. A number of full-wave techniques for the analysis of planar transmission lines and printed circuits are available. These methods include the Finite Element Method (FEM), the Finite-Difference Time-Domain (FDTD) method, the Method of Line (MoL), the Transmission Line Matrix (TLM) and the Method of Moments (MoM).

A rigorous full-wave solution using a numerical method always requires a relative high degree of numerical effort. As a result, these numerical methods lead to long computation time. Based on the criteria of computational efficiency and accuracy, the MoM is a popular and rigorous full-wave method for the analysis of planar transmission lines and associated passive circuits devices, and it can be formulated either in the spatial domain or in the spectral domain. The two approaches are physically equivalent, but they are different in the mathematical implementation. For planar structures, the spectral domain approach is simpler and efficient because the partial differential equations are reduced to ordinary ones, with respect to the direction normal to the substrate surface. These equations are then solved using matrix techniques.

Accurate results based on the MoM depend totally on the evaluation of the matrix elements. The computation of these matrix elements is the most time consuming part in the MoM because the matrix elements are expressed in terms of infinite integrals. In practical cases, the infinite integrals have to be truncated at sufficiently large upper limits such that convergence is attained. This requires lengthy computation time for the numerical integration of the impedance matrix. In order to improve the computational efficiency associated with calculating these matrix elements, several authors [68]-[70] use a hybrid method, combining the spectral and spatial domains MoM, which employs the spatial domain technique for the asymptotic solution of the spectral domain integral. The hybrid method, to calculate the spatial domain portion, requires additional computations of the same structure in a homogeneous medium with an average dielectric constant immediately above and below the conductor. This technique, although more efficient than the original SDA, still requires many numerical integrations.

A new approach is proposed here which promises to alleviate these problems. Asymptotic extraction techniques are applied to convert the slowly converging impedance matrix elements used in the SDA into the sum of a rapidly converging term and a slowly converging term (asymptotic part of impedance matrix elements). The integrand of the latter term consists of the product of the asymptotic form of the Green's function and basis functions. Instead of using a "brute force" numerical integration, we propose to use the closed-form or analytical solution for calculating the asymptotic part of the impedance matrix elements in the spectral domain. For some specific structures, the latter term (asymptotic part of impedance matrix elements) can be integrated analytically by introducing appropriate basis functions. In the present work, we are concerned with finding the closed-form or analytical solution of the asymptotic part of the impedance matrix elements for planar transmission lines, and printed circuits such as microstrip dipoles, asymmetric gap discontinuities, and arbitrary shapes of printed circuits.

4.2 Research Objectives

The main goal of the proposed research is to improve the computational efficiency of the spectral domain moments method using the closed-form solution for the asymptotic part of the impedance matrix elements. The application and emphasis will be placed on the following structures:

- Planar Transmission Line (Fig. 4.1)

    - Multilayer single microstrip line(Fig. 4.1a)

    - Multilayer slotline (Fig. 4.1b)

    - Multilayer coupled microstrip line (Fig. 4.1c)

    - Multilayer coplanar waveguide (Fig. 4.1d)

- Discontinuities in Microstrip Line

    - Symmetric gap (Fig. 4.2a)

    - Asymmetric gap (Fig. 4.2b)

- Microstrip Dipoles (Fig. 4.3a)

- Radar Cross Section Problem of Microstrip patch (Fig. 4.3b)

The analytical asymptotic extraction method developed in this work, especially the formulation using roof-top basis functions, can be applied for the analysis of any arbitrary planar circuit of the form shown in Fig. 4.4.

4.3 Asymptotic Extraction Technique

4.3.1 Planar transmission line

In analyzing planar transmission lines shown in Fig. 4.1, an integral equation in the spectral domain is formulated by using the Green's function and the basis functions

74

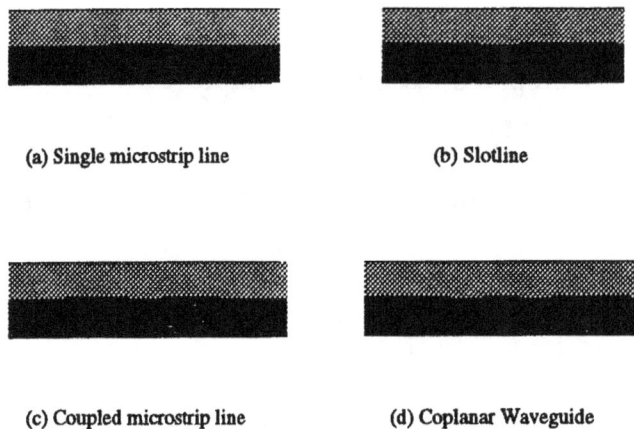

(a) Single microstrip line

(b) Slotline

(c) Coupled microstrip line

(d) Coplanar Waveguide

Fig. 4.1: Cross sectional geometries of planar microstrip transmission

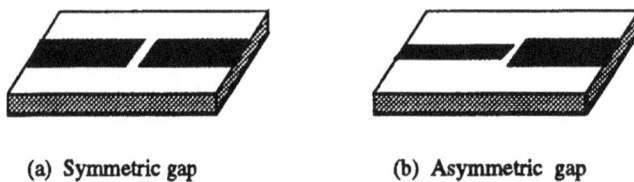

(a) Symmetric gap

(b) Asymmetric gap

Fig. 4.2: Discontinuities in Microstrip Line

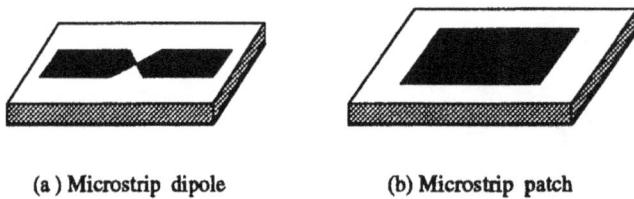

(a) Microstrip dipole

(b) Microstrip patch

Fig. 4.3: Microstrip dipole and patch

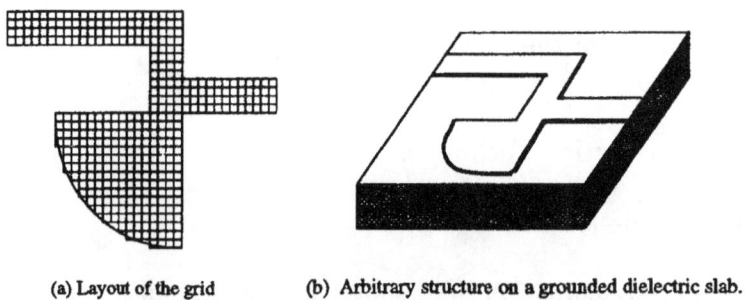

(a) Layout of the grid

(b) Arbitrary structure on a grounded dielectric slab.

Fig. 4.4: Arbitrary planar circuit and its grid

with the unknown current coefficients. Two-dimensional Fourier transforms are carried out along the longitudinal and transverse directions ($\beta - \alpha$ plane in the spectral domain). In the spectral domain formulation, convolution-type integral equations of the tangential electric fields at the interface on the conductor strip lines can be derived and typically are represented by algebraic equations of the form [71].

$$\frac{j}{\omega\epsilon_0}\left[\tilde{J}_{zm}(\alpha)\tilde{G}_{zz}(\alpha,\beta) + \tilde{J}_{xn}(\alpha)\tilde{G}_{zx}(\alpha,\beta)\right] = \tilde{E}_z(\alpha) \tag{4.1}$$

$$\frac{j}{\omega\epsilon_0}\left[\tilde{J}_{zm}(\alpha)\tilde{G}_{xz}(\alpha,\beta) + \tilde{J}_{xn}(\alpha)\tilde{G}_{xx}(\alpha,\beta)\right] = \tilde{E}_x(\alpha) \tag{4.2}$$

where $\tilde{G}_{zz}$, $\tilde{G}_{zx}$ and $\tilde{G}_{xx}$ are dyadic Green's functions for the structure of interest, and $\tilde{J}_{zm}$ and $\tilde{J}_{xn}$ are Fourier transforms of the basis functions.

The main objective is to solve (4.1) and (4.2) for the phase constant β which is representative of the propagation and dispersive characteristics of the transmission line. We will outline here a method that can be used to accomplish this. To solve for β [and $\epsilon_{reff}(f)$], the current density distribution on the strips is expanded in terms of a known set of basis functions with unknown coefficients. Application of Galerkin's method and Parseval's theorem [72] converts (4.1) and (4.2) into a system of algebraic linear equations written in matrix form as

$$\begin{bmatrix}[Z_{mm}] & [Z_{mn}]\\ [Z_{nm}] & [Z_{nn}]\end{bmatrix}\begin{bmatrix}[a_m]\\ [b_n]\end{bmatrix} = \begin{bmatrix}[0]\\ [0]\end{bmatrix} \quad \begin{array}{l}m = 0,1,2,\cdots,M-1\\ n = 0,1,2,\cdots,N-1\end{array} \tag{4.3}$$

The matrix elements are defined as [71];

$$Z_{mm} = \int_{-\infty}^{\infty} \tilde{J}_{zm}(\alpha)\tilde{G}_{zz}(\alpha,\beta)\tilde{J}_{zm}^*(\alpha)\,d\alpha \tag{4.4}$$

$$Z_{mn} = \int_{-\infty}^{\infty} \tilde{J}_{xn}(\alpha)\tilde{G}_{zx}(\alpha,\beta)\tilde{J}_{zm}^*(\alpha)\,d\alpha \tag{4.5}$$

$$Z_{nm} = \int_{-\infty}^{\infty} \tilde{J}_{zm}(\alpha)\tilde{G}_{xz}(\alpha,\beta)\tilde{J}_{xn}^*(\alpha)\,d\alpha \tag{4.6}$$

$$Z_{nn} = \int_{-\infty}^{\infty} \tilde{J}_{xn}(\alpha)\tilde{G}_{xx}(\alpha,\beta)\tilde{J}_{xn}^*(\alpha)\,d\alpha \tag{4.7}$$

To find the propagation constant β, which determines the dispersive characteristics of the transmission line, a nonlinear equation solver can be used. The solution is based on the vanishing of the determinant in (4.3).

The evaluation of the impedance matrix elements of (4.3), which are represented by (4.4)-(4.7), includes the integral over infinite limits which is very time consuming because of the slow convergence and the highly oscillating behavior of the integrands. In a practical approach, the infinite limits are truncated at sufficiently large values of α to obtain accurate results. The propagation constants of the transmission lines are very sensitive to where the integration is truncated and how it is subsequently discretized. This is a significant limiting factor in the integration. To overcome this problem, the spectral domain approach is usually employed in conjunction with an acceleration technique referred to as the *asymptotic extraction* [68]. The asymptotic part of the Green's function is subtracted and added from the original Green's function and the integral of (4.4), for example, is split into two parts

$$Z_{mm} = \int_0^\infty \tilde{J}_{zm}(\alpha) \left[\tilde{G}_{zz}(\alpha, \beta) - \tilde{G}_{zz}^\infty(\alpha, \beta) \right] \tilde{J}_{zm}(\alpha)^* \, d\alpha$$

$$+ \int_0^\infty \tilde{J}_{zm}(\alpha) \tilde{G}_{zz}^\infty(\alpha, \beta) \tilde{J}_{zm}(\alpha)^* \, d\alpha \tag{4.8}$$

which can be approximated by

$$Z_{mm} \simeq \int_0^{\alpha_u} \tilde{J}_{zm}(\alpha) \left[\tilde{G}_{zz}(\alpha, \beta) - \tilde{G}_{zz}^\infty(\alpha, \beta) \right] \tilde{J}_{zm}(\alpha)^* \, d\alpha$$

$$+ \int_0^\infty \tilde{J}_{zm}(\alpha) \tilde{G}_{zz}^\infty(\alpha, \beta) \tilde{J}_{zm}(\alpha)^* \, d\alpha \tag{4.9}$$

All other impedance matrix elements Z_{mn}, Z_{nm}, and Z_{nn} of (4.5)-(4.7) are treated in a similar manner.

Subtraction of the asymptotic terms from the Green's functions makes the integrands of the first integrals of (4.8) and (4.9) decay faster to zero for large α, so the integrals can be truncated at an upper limit α_u which can then be numerically calculated. The numerical calculations of the first integrals of (4.8) and (4.9) have been addressed by other literatures [73], [74], and it is not the subject of this research.

As an initial step to investigate the asymptotic closed-form extractions for the impedance integral, we examine the asymptotic behavior of the Green's function, with respect to α for large α. For planar transmission problems, we found that the asymptotic Green's functions have following simple forms [75].

$$\tilde{G}_{zz}^{\infty} \propto \frac{1}{|\alpha|} \tag{4.10}$$

$$\tilde{G}_{xx}^{\infty} \propto |\alpha| \tag{4.11}$$

$$\tilde{G}_{xz}^{\infty} \propto \text{sgn}(\alpha) \tag{4.12}$$

Using this asymptotic behavior of the Green's function and the appropriately chosen basis functions, we can solve in closed-form for the second term of the integral in (4.9). In transmission line problems, Chebyshev polynomial basis functions with edge condition are suitable, which in some specific transmission line structures allow closed-form solutions in the asymptotic part of impedance matrix.

We already found two types of asymptotic closed-form solutions which can be applied to the single microstrip line, slotline, coupled microstrip line, CPW, and the scattering problem of a finite array of infinite strips(with or without the grounded dielectric slab).

4.3.2 Asymptotic extraction technique of microstrip printed circuits

The propagation of electromagnetic waves in a grounded dielectric slab has numerous applications in printed antenna technology and in the analysis of microwave and millimeter wave integrated circuits. For the accurate analysis of microstrip dipoles and circuits based on the Method of Moments (MoM), a crucial step is precise evaluation of the impedance matrix elements which contain the integration of Sommerfeld-type integrals. For the accurate and efficient computation of these impedance matrix elements, numerous researchers [76]-[81] have extensively studied the problem to obtain the dielectric slab Sommerfeld-type Green's function with approximate closed-form in the spatial domain.

For planar structures, the spectral domain approach(SDA) is popular and efficient method. However, in the spectral domain, filling the impedance matrix elements is the most time consuming part in the MoM because the matrix elements are expressed in terms of infinite double integrals and their integrands exhibit slow convergence and highly oscillating behavior. In order to improve the computational efficiency associated with calculating these matrix elements, this research uses the asymptotic extraction technique for the evaluation of the impedance matrix element in the spectral domain. And then, for each planar geometry, the infinite double integral of the asymptotic impedance matrix element is transformed into a finite one-dimensional integral with an analytical technique.

In the planar circuits, such as microstrip dipoles, and discontinuities in microstrip lines and arbitrarily shaped planar circuits, a typical impedance matrix elements of moment method can be written as [68]

$$\overline{\overline{Z}}_{mn} = -\int_x \int_y \int_{x_0} \int_{y_0} \int_{k_x=-\infty}^{\infty} \int_{k_y=-\infty}^{\infty} \vec{J}_m(x_0,y_0)\overline{\overline{\tilde{G}}}(k_x,k_y)\vec{J}_n(x,y)$$

$$\times e^{jk_x(x-x_0)}e^{jk_y(y-y_0)}\, dk_x dk_y dy_0 dx_0 dy dx \tag{4.13}$$

Each impedance matrix element in the spatial domain consists of six-fold integral: Of these integrations, two are used to convert Sommerfeld-type Green's function into the spatial domain, two are convolution integrals, and the remaining two are the inner products. These six-fold type integrals are numerically intensive in the spatial domain. In planar structures, the spectral domain approach is preferred and the matrix elements can be rewritten as [30],[29]

$$\overline{\overline{Z}}_{mn} = -\frac{1}{4\pi^2}\int_{-\infty}^{\infty}\int_{-\infty}^{\infty} \vec{\tilde{J}}_m(k_x,k_y)\overline{\overline{\tilde{G}}}(k_x,k_y)\vec{\tilde{J}}_n^*(k_x,k_y)\, dk_x dk_y \tag{4.14}$$

Numerically computing the double integration is very inefficient since the number of computations is proportional to the square of the number of terms taken for a single integration. To improve the computation efficiency, we use the asymptotic extraction technique in which the asymptotic part of the Green's function is subtracted and added from the original Green's function as [68]

$$\overline{\overline{Z}}_{mn} = -\frac{1}{4\pi^2} \int_{-\infty}^{\infty} \int_{-\infty}^{\infty} \vec{\tilde{J}}_m(k_x, k_y)[\overline{\overline{\tilde{G}}}(k_x, k_y) - \overline{\overline{\tilde{G}}}^{\infty}(k_x, k_y)]\vec{\tilde{J}}_n^*(k_x, k_y) \, dk_x dk_y$$

$$-\frac{1}{4\pi^2} \int_{-\infty}^{\infty} \int_{-\infty}^{\infty} \vec{\tilde{J}}_m(k_x, k_y)\overline{\overline{\tilde{G}}}^{\infty}(k_x, k_y)\vec{\tilde{J}}_n^*(k_x, k_y) \, dk_x dk_y \qquad (4.15)$$

The first integral decays rapidly to zero as β becomes large. In fact, the double infinite integral can be truncated at a finite upper limit β^u after transforming the Cartesian coordinates (k_x, k_y) into the polar coordinate (β, ϕ). Thus the first integral can be evaluated by any suitable numerical technique with negligible error and the second integral can be solved analytically. To derive the closed-form solution of the second integral in (4.15), first we need to find the asymptotic behavior of Green's function. The asymptotic behaviors of the dielectric slab Green's function for large k_x and k_y are given by

$$\tilde{G}_{xx}^{\infty}(k_x, k_y) = -j\frac{Z_0}{k_0}\left\{\frac{k_0^2}{2\beta} - \frac{k_x^2}{(\epsilon_r + 1)\beta}\right\} \qquad (4.16)$$

$$\tilde{G}_{yy}^{\infty}(k_x, k_y) = -j\frac{Z_0}{k_0}\left\{\frac{k_0^2}{2\beta} - \frac{k_y^2}{(\epsilon_r + 1)\beta}\right\} \qquad (4.17)$$

$$\tilde{G}_{xy}^{\infty}(k_x, k_y) = \tilde{G}_{yx}^{\infty}(k_x, k_y) = j\frac{Z_0}{k_0}\frac{k_x k_y}{(\epsilon_r + 1)\beta} \qquad (4.18)$$

where $\beta = \sqrt{k_x^2 + k_y^2}$.

The present form of the above asymptotic Green's functions can be also applicable to generalized multilayer planar structures. This is based on the fact that the multilayer planar structures are asymptotically equivalent to two layers structures by eliminating layers farther away from the source and extending each adjacent layer of the new structure to infinity. The asymptotic limit of Green's functions depends on the thickness of the layer adjacent to the current element. For thicker layers, we can expect that the convergence of the asymptotic behavior is better.

With this asymptotic behavior of the Green's functions given in (4.16)-(4.18) and properly chosen basis functions, the asymptotic parts of the matrix elements in (4.15) will be solved analytically for the three-dimensional planar structures including microstrip dipoles, asymmetric gap discontinuities, and arbitrarily shaped planar circuits.

4.4   Results of Planar Transmission Lines

Two different types of closed-form solutions have been obtained for the asymptotic part of the impedance matrix elements in open and coupled microstrip lines. Both of the formulas were solved by using Chebyshev polynomials basis functions with appropriate edge condition and limiting behavior of the Green's function. To check the validity of our improved computational method using the asymptotic closed-form solution, we briefly present the results of the propagation constants (effective dielectric constant) for a single microstrip line and for a symmetric coupled microstrip line.

4.4.1   Single microstrip line

The second term of the integral in (4.9) in a single microstrip line in a multi-layer structure can be written in closed-form as

$$I_{mn} = \int_0^\infty \frac{J_m\left(\frac{w\alpha}{2}\right) J_n\left(\frac{w\alpha}{2}\right)}{\alpha} \, d\alpha = \frac{2}{\pi} \frac{\sin\left(\frac{(m-n)\pi}{2}\right)}{m^2 - n^2} \tag{4.19}$$

provided that $Re(m + n) > 0$, $w > 0$, and where $J_m(\alpha)$ is the Bessel function of the first kind.

All other asymptotic matrix elements are treated in a similar manner. However, when the basis and weighting functions are simultaneously zero order Bessel functions (as for matrix element Z_{00}), the closed-form solution of (4.19) cannot be applied because $Re(m+n) = 0$. In that case, another approach has to be used to evaluate the second term of (4.9) for $m = n = 0$. To evaluate Z_{00}, we go back and rewrite (4.9) in its original form without subtracting and adding the asymptotic form of the Green's function. Doing this, we can write Z_{00} as

$$Z_{00} = \int_0^\infty \tilde{J}_{z0}(\alpha)\tilde{G}_{zz}(\alpha,\,\beta)\tilde{J}_{z0}^*(\alpha)\,d\alpha \qquad (4.20)$$

which can be approximated by

$$Z_{00} \simeq \int_0^{\alpha_u} \left[\tilde{J}_{z0}(\alpha)\tilde{G}_{zz}(\alpha,\,\beta)\tilde{J}_{z0}^*(\alpha)\right]\,d\alpha$$

$$+ \int_{\alpha_u}^\infty \left[\tilde{J}_{z0}(\alpha)\tilde{G}_{zz}^\infty(\alpha,\,\beta)\tilde{J}_{z0}^*(\alpha)\right]\,d\alpha \qquad (4.21)$$

The second integral of (4.21) can be rewritten in the following form

$$I_{00} = \int_{\alpha_u}^\infty \frac{J_0(\frac{w\alpha}{2})\,J_0(\frac{w\alpha}{2})}{\alpha}\,d\alpha \qquad (4.22)$$

where $J_0(\alpha)$ is the Bessel function of the first kind of order 0.

The integral of (4.22) can be solved in terms of series as follows [75]

$$I_{00} = \sum_{k=0}^\infty \frac{(-1)^k \left(\frac{w\alpha_u}{2}\right)^{(2k+2)}(2k+1)!}{(k+1)^2((k+1)!)^2(k!)^2 2^{(2k+2)}} - \gamma - \ln\left(\frac{w\alpha_u}{4}\right) \qquad (4.23)$$

which converges uniformly for all real values except for $\alpha_u = 0$.

Using the asymptotic closed-form solutions (4.19) and (4.23), the effective dielectric constants of an open microstrip line for $w/h=1$ ($\epsilon_r = 8$, $\mu_r = 1$) were calculated and plotted in Fig. 4.5. We compared our results with those of Kobayashi et al. [82], [83] using the conventional SDA, and Shin et al. [84] using variational conformal mapping. The agreement is good, although there are slight discrepancies which cannot be discerned in Fig. 4.5.

Table 4.1 illustrates a comparison of the computational time between the conventional SDA without asymptotic extraction technique and the proposed method for the calculations of effective dielectric constant for $w/h = 1$ and $w/h = 0.1$ (

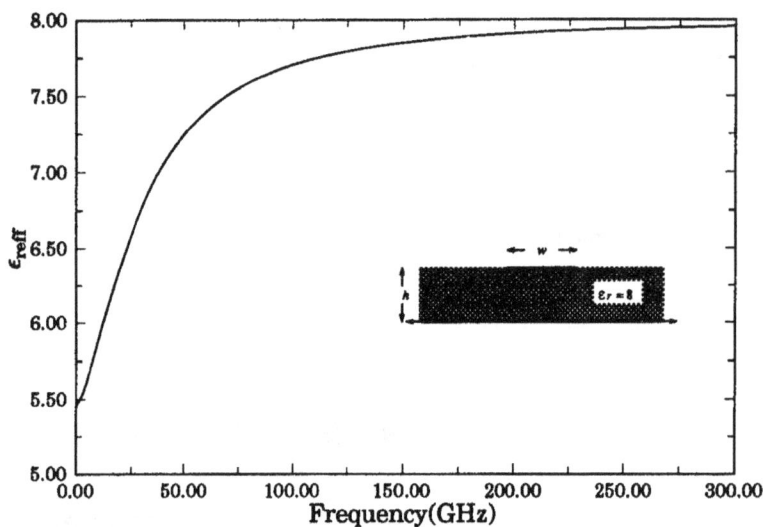

Fig. 4.5: Effective dielectric constants versus frequency (w/h=1, $\epsilon_r = 8$, μ_r=1).

Table 4.1: Computer Time on a SUN SPARC station for the Calculation of the Effective
Dielectric Constant with Two Different Techniques in the single microstrip line ($h/\lambda_0 =$
0.1 $\epsilon_r = 8$, $\mu_r = 1$)

	SDA without asymptotic technique [a] ($\Delta\alpha = 0.01, \alpha_u = 1000$)	Proposed Method[b] ($\Delta\alpha = 0.01, \alpha_u = 30$)	Numbers of Iteration both a and b	Computational Efficiency ($\frac{a}{b}$)
$\frac{w}{h} = 1$ (ϵ_{reff})	201.360 seconds (6.7576)	7.6 seconds (6.7572)	7	26.5
$\frac{w}{h} = 0.1$ (ϵ_{reff})	288.060 seconds (5.7719)	5.63 seconds (5.7697)	6	51.165

$h/\lambda_0 = 0.1$). The initial value to find the real roots of the determinant of the system matrix is very important in Muller's root finding method. If we have a sufficiently good initial guess, the method converges rapidly. For both techniques, the quasi-TEM (p. 450 in [85]) effective dielectric constant was used as the initial value. Starting with this initial trial solution, the results shown in Table 4.1 converge, with an accuracy of 10^{-4}, after the seventh iteration for $w/h = 1$ and after the sixth iteration for $w/h = 0.1$. The conventional SDA requires a significantly greater amount of computer time than the proposed method. As shown in Table 4.1, the improved method reduces the computational time by 26 times (for $w/h = 1$) and 51 times (for $w/h = 0.1$) than the conventional SDA. For $w/h = 0.1$, we notice that the effective dielectric constant 5.7719, obtained from the conventional SDA in Table 4.1, is slightly different from the values of 5.765, obtained form [83]. This indicates that for the narrow strip ($w/h = 0.1$) the upper limit α_u needs to be increased in order to obtain the same values as in [83]. The closed-form formula in (4.19) can also be used to solve the slotline problem.

4.4.2 Symmetric coupled microstrip line

To solve the dispersion characteristics of symmetric coupled microstrip lines, we used the even and odd modes method. In this method wave propagation along a coupled pair of strip conductor lines is expressed in terms of the even and odd modes by placing a magnetic wall and a electric wall, respectively, at the plane of symmetry. Because the center of strip conductor lines is now situated at $x = \pm(s+w)/2$ instead of at the origin, the currents in the spectral domain must be multiplied by $e^{\pm j(s+w)/2}$. Using the even and odd modes method, the asymptotic part of the impedance matrix elements in (4.9) in the symmetric coupled microstrip line can be written as [86]

$$I_{mn} = \int_0^\infty \frac{J_m\left(\frac{w\alpha}{2}\right) J_n\left(\frac{w\alpha}{2}\right)}{\alpha} [1 \pm \cos((s+w)\alpha)]\, d\alpha, \quad \begin{cases} + & \text{if even mode} \\ - & \text{if odd mode} \end{cases} \quad (4.24)$$

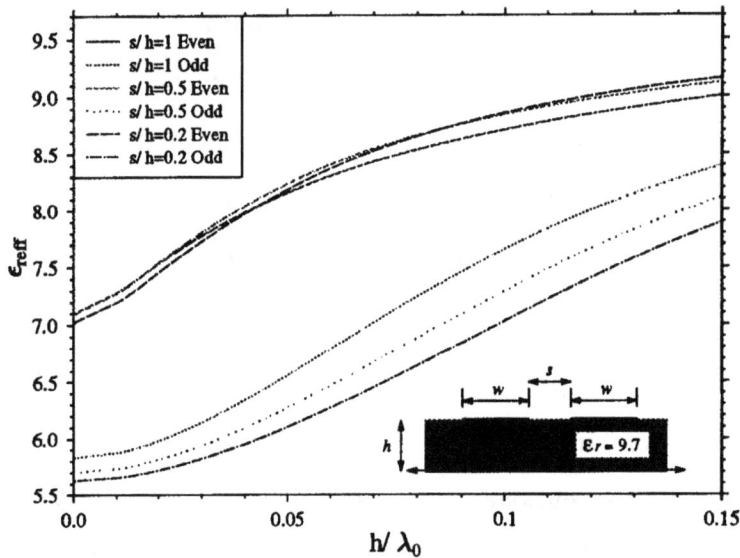

Fig. 4.6: Effective dielectric constants versus h/λ_0 for $s/h = 1$, $s/h = 0.5$, $s/h = 0.2$ ($w/h=1$, $\epsilon_r = 9.7$, $\mu_r=1$).

The first term of the integral in (4.24) is carried out by using (4.19) and (4.23). The second term of the integral in (4.24), after tedious mathematical manipulations, is also solved in the closed-form as [86]

$$\int_0^\infty \frac{J_m\left(\frac{w\alpha}{2}\right) J_n\left(\frac{w\alpha}{2}\right)}{\alpha} \cos((s+w)\alpha)\,d\alpha = \left(\frac{w}{4(s+w)}\right)^{m+n} \cos\left(\frac{m+n}{2}\pi\right)$$

$$\times \left\{ \sum_{p=0}^\infty \frac{1}{(m+n+2p)} \left(\frac{m+n+2p}{p}\right) \left(\frac{w}{4(s+w)}\right)^{2p} \frac{(m+n+2p)!}{(m+p)!(n+p)!} \right\} \quad (4.25)$$

provided that $\left(\frac{w}{s+w}\right) < 1$.

The numerical results obtained by the proposed method using the closed-form solutions of (4.19), (4.23) and (4.25) are plotted in Fig. 4.6 for s/h=0.2, 0.5, 1 (ϵ_{reff}=9.7, w/h=1). We compared our results with those of Kowalski $et\ al.$ [87] using the conventional SDA. They are in good agreement with each other.

To compare the computational time between the conventional SDA without

Table 4.2: Computer Time on a SUN SPARC station for the Calculation of the Effective Dielectric Constant with Two Different Techniques in the Coupled Microstrip Line ($h/\lambda_0 = 0.1$, $\epsilon_r = 9.7$, $\mu_r = 1$)

	SDA without asymptotic technique[a] ($\Delta\alpha = 0.01, \alpha_u = 1000$)	Proposed Method[b] ($\Delta\alpha = 0.01, \alpha_u = 30$)	Numbers of Iteration both a and b	Computational Efficiency ($\frac{a}{b}$)
Even mode (ϵ_{reff})	247.91 seconds (8.7128)	8.15 seconds (8.7077)	7	30.41
Odd mode (ϵ_{reff})	247.90 seconds (7.6451)	8.15 seconds (7.6511)	7	30.41

asymptotic extraction technique and the proposed method, the effective dielectric constants of the even mode and odd mode are listed in Table 4.2, for ϵ_{reff}=9.7, $w/h = 1$, $s/h = 1$ and h/λ_0=0.1. As shown in Table 4.2, the proposed method reduces the computational time by 30 times for both the even and odd modes compared to the conventional SDA.

Formulas (4.19), (4.23), and (4.25) can also be applied to the CPW problem. Moreover three formulas, similar to those of (4.19), (4.23) and (4.25), can also be applied to the scattering by a finite array of strips (with or without the dielectric grounded slab).

4.5 Results of Microstrip Printed Circuits

For the evaluation of the impedance matrix elements in planar circuits, we present the analytical transformation technique. Using this method, the infinite double integral of the asymptotic impedance matrix elements is transformed into a finite one-dimensional integral. It is interesting that the finite one-dimensional integral is especially more efficient in the cases of the smaller size of basis functions and larger lateral separations between the basis and testing functions, which is a pathological case of the conventional spectral domain analysis. The formula presented in this work produces accurate and efficient results to evaluate the asymptotic part of impedance matrix without limitations for the analysis of planar circuits. This results into a dramatic improvement of the computation time for evaluating the overall impedance matrix elements for the following specific geometries; microstrip dipole, symmetric and asymmetric gap, and scattering of rectangular microstrip patch.

4.5.1 Microstrip dipole

To describe the anticipated currents along the electrically narrow microstrip lines, triangular subdomain basis functions with edge condition are used. The edge condition with the square-root weighting function describes well the singular behavior of the longitudinal current densities toward the edges of the strips. Also the transverse current is assumed to be zero, because the width of the strip is generally considered very thin [30]. Thus, only the matrix $[Z_{mn}^{xx}]$ involving the Green's function $\tilde{G}_{xx}$ needs to be evaluated.

The longitudinal current densities of the triangular basis functions with edge condition are denoted by $J_{xm}(x,y)$, where $J_{xm}(x,y)$ is defined as

$$J_{xm}(x,y) = \frac{1 - \frac{|x-x_m|}{L}}{\sqrt{1 - \left(\frac{2y}{W}\right)^2}}, \qquad \left|\frac{2y}{W}\right| < 1, \; \left|\frac{x - x_m}{L}\right| < 1 \qquad (4.26)$$

where W is the width of the strip, and L is the half-length of the basis function.

Substituting (4.16) and the spectral domain representation of (4.26) into the second integral of (4.15), the asymptotic part of impedance matrix $[Z_{mn}^{xx}]$ is written as [88]

$$Z_{mn}^{xx^{Asy}} = -j\frac{1}{\pi^2}\frac{Z_0}{k_0}\left(2\pi\frac{W}{L}\right)^2\left\{-\frac{k_0^2}{2}I_{mn}^a + \frac{1}{(\epsilon_r + 1)}I_{mn}^b\right\} \qquad (4.27)$$

with

$$I_{mn}^a = \int_0^\infty \int_0^\infty \frac{\cos(x_s k_x)}{\sqrt{k_x^2 + k_y^2}}\frac{\sin^4\left(k_x\frac{L}{2}\right)}{k_x^4}\left[J_0\left(k_y\frac{W}{2}\right)\right]^2 dk_x dk_y \qquad (4.28)$$

$$I_{mn}^b = \int_0^\infty \int_0^\infty \frac{\cos(x_s k_x)}{\sqrt{k_x^2 + k_y^2}}\frac{\sin^4\left(k_x\frac{L}{2}\right)}{k_x^2}\left[J_0\left(k_y\frac{W}{2}\right)\right]^2 dk_x dk_y \qquad (4.29)$$

where the even and odd properties of the integrand are used to reduce the integration range in (4.28) and (4.29), and x_s is defined as $|x_m - x_n|$.

Using the analytical technique presented in [88], the infinite double integrals of (4.28) and (4.29) are transformed into finite one-dimensional integrals as

$$I_{mn}^a = \frac{1}{2} \int_{-2L}^{2L} A(\chi - x_s) \cdot \Im_a(\chi) \, d\chi \qquad (4.30)$$

$$I_{mn}^b = \frac{1}{2} \int_{-2L}^{2L} A(\chi - x_s) \cdot \Im_b(\chi) \, d\chi \qquad (4.31)$$

where $A(\chi - x_s)$, $\Im_a(\chi)$, and $\Im_b(\chi)$ are defined in [88].

To check the validity of the newly derived formulas, the input impedance of a center-fed microstrip dipole is obtained by using (4.30) and (4.31) for efficiently evaluating a poorly convergent asymptotic impedance matrix in (4.27). In microstrop dipole, the matrix $[Z_{mn}^{xx}]$ are determined by a Galerkin moment method in the spectral domain based on the acceleration technique in (4.15). And the impressed voltage is represented by a delta-gap generator at the center-fed point [89],[90].

The microstrip dipole on the dielectric slab has length H and width $W = 0.01\lambda_0$ with a relative permittivity $\epsilon_r = 3.25$ and a substrate thickness $d = 0.0796\lambda_0$. The values of the input impedance, computed by using the proposed method, are compared with those of the conventional method in Fig. 4.7(a), in which the conventional method gives the same results of Marin et $al.$ [89]. The conventional method was calculated by using three expansion modes [Piecewise Sinusoidal(PWS) basis with edge condition along the $\hat{y}$ direction] in the spectral domain. To reach the results of the conventional method, the proposed method used seven expansion modes. There seems to be a good agreement between the methods; however a slight discrepancy is observed in the peak impedance regions in Fig. 4.7(a).

In the proposed method, the first resonance occurs at $H \simeq 0.29\lambda_0$, and the second at $H \simeq 0.57\lambda_0$, which is similar to the results indicated in [89]. By increasing the number of basis functions, the first resonance($H \simeq 0.29\lambda_0$) is almost unchanged. In

88

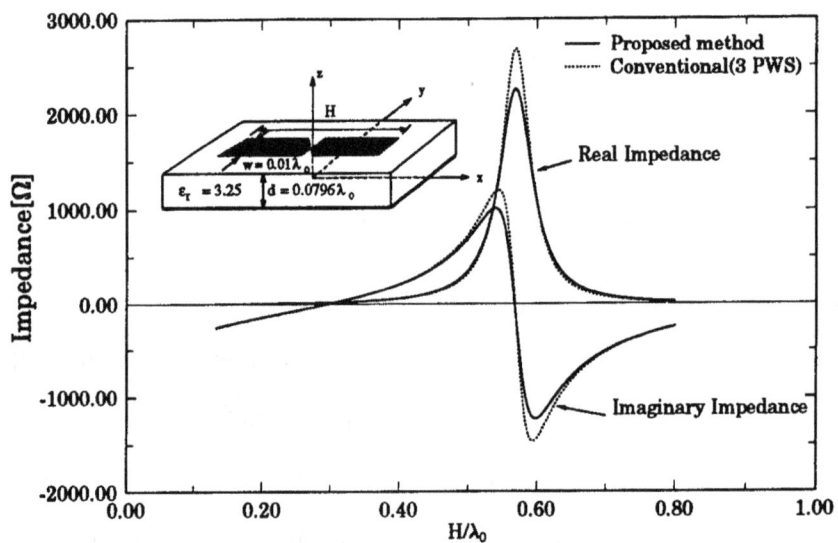

(a)

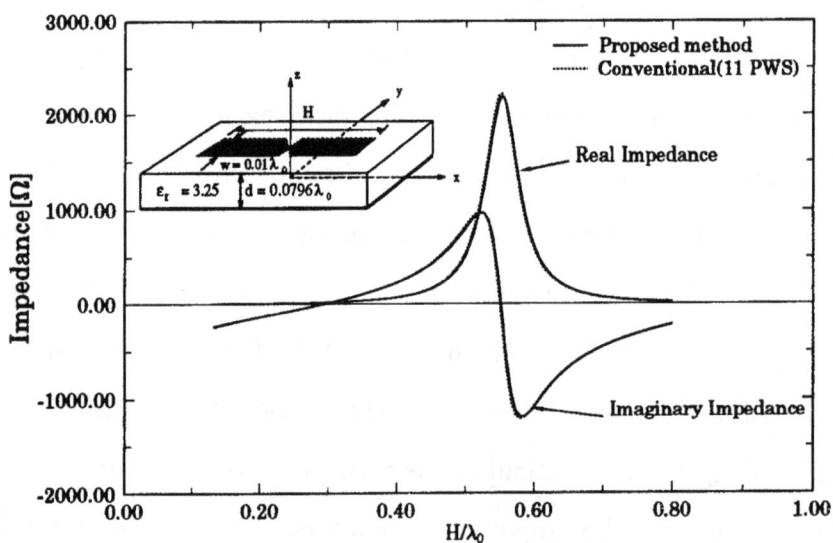

(b)

Fig. 4.7: Input impedance of center-fed microstrip antenna

contrast to the first resonance, we have found the second resonance to be somewhat sensitive if a small number of mode expansions is used. This indicates that more basis functions are needed for this structure. The convergence of the second resonance is achieved by increasing the number of the basis functions to thirteen for the proposed method and to eleven basis functions for the conventional method, respectively. In this case, it is found that both methods indicate the second resonance at $H \simeq 0.55\lambda_0$, with the first resonance remaining almost unchanged. The input impedances of both results versus normalized length (H/λ_0) are plotted in Fig. 4.7(b). It can be seen that the agreement with each method is quite good: the agreement becomes better by increasing the number of basis functions.

To calculate the impedance matrix $[Z^{xx}_{mn}]$ using the proposed method, the upper limit β^u of the first integral in (4.15) was set to $50 \cdot k_0$. A further increase in the upper limit β^u on the first integral in (4.15) did not lead to any substantial change of the input impedance. Meanwhile, for a direct calculation of the impedance matrix of (4.14)[using the conventional method], the upper limit $\beta^u = 400 \cdot k_0$ was used. For comparison of the overall computational efficiency, the average computation time was calculated for the two methods used to obtain the results of Figures 4.7(a) and (b). The overall computation time of the proposed method(with seven expansion modes) is 9 times faster than that of the conventional method(with three expansion modes) for the results of Fig. 4.7(a) and 12 times faster for the results of Fig. 4.7(b). Also the accuracy of the proposed method is quite comparable with that of the conventional method.

4.5.2 Asymmetric gap discontinuities

Accurate analysis of asymmetric gap discontinuities is a computationally more time-intensive problem than those of symmetric gaps. This is due to the fact that if the width of two electrical narrow strip lines is mismatched, the integrand of each interaction matrix of an asymmetric gap is more highly oscillatory compared to those of a symmetric gap. In order to speed up the execution time, this part presents an analytical technique for solving the asymptotic part of impedance matrix of the general asymmetric gap problems [91], [92].

The asymmetric gap in open microstrip structures is subject to radiation at discontinuities in the form of either space or surface wave. Full-wave solution is needed to take into account such phenomena. An asymmetric gap discontinuity on a grounded dielectric substrates is depicted in Fig. 4.8. The current expansions for an asymmetric gap are shown in Fig. 4.9. Unknown current densities are modeled by the standing and traveling wave modes and the expansion function with triangular edge mode basis. The cosine wave mode starts at a quarter wavelength away from the gap discontinuities to satisfy the boundary condition at the end of the line. And the expansion function covers at least quarter wavelength from gap discontinuities.

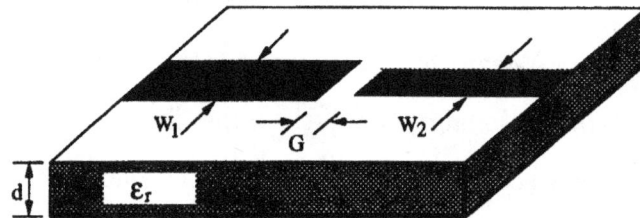

Fig. 4.8: Geometry of asymmetric gap discontinuities

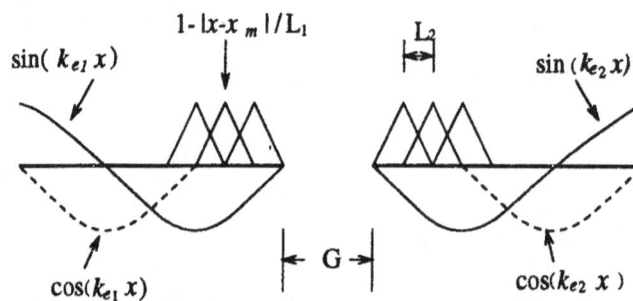

Fig. 4.9: Current expansions on the asymmetric gap

Electric field boundary condition is imposed on the conductor strip. And the electric field integral equation is then converted into a matrix system by multiplying by a testing function and integrating the inner product over the support of this function. As a result, the matrix system for an asymmetric gap shown in Fig. 4.8 is formulated as [91], [92]

$$
\begin{bmatrix} [Z_{mn}] & [Z_{mc}+jZ_{ms}] & [Z_{mq}] & [Z_{mtc}+jZ_{mts}] \\ [Z_{pn}] & [Z_{pc}+jZ_{ps}] & [Z_{pq}] & [Z_{ptc}+jZ_{pts}] \end{bmatrix} \begin{bmatrix} [I_n] \\ -R \\ [I_q] \\ T \end{bmatrix} = \begin{bmatrix} [-Z_{mc}+jZ_{ms}] \\ [-Z_{pc}+jZ_{ps}] \end{bmatrix} \tag{4.32}
$$

Each submatrix in (4.32) represents a set of mutual interactions between the test functions and the basis functions. Their respective mathematical representations and asymptctic forms of each submatrix are presented in [92]. Since the integrand of the double infinite integration in each submatrix element converges slowly with a highly oscillatory behavior, a numerical integration is extremely laborious to evaluate. Thus, the analytical procedures of the asymptotic part of impedance matrix, if applicable, can considerably reduce the computational effort involving Sommerfeld-type integration.

The asymptotic submatrices of Z_{mn} and Z_{pq} have similar formulas in (4.28) and (4.29). Thus, their analytical results have similar formulas as in (4.30) and (4.31). The asymptotic part of submatrices Z_{mq} and Z_{pn} has the following two general functions given by [91], [92]

$$
I_{mq}^a = \int_0^\infty \int_0^\infty \frac{\cos(d_s k_x)}{\sqrt{k_x^2+k_y^2}} \frac{\sin^2(k_x L_1/2)}{k_x^2} \frac{\sin^2(k_x L_2/2)}{k_x^2} J_0\left(k_y\frac{W_1}{2}\right) J_0\left(k_y\frac{W_2}{2}\right) dk_x\, dk_y \tag{4.33}
$$

$$
I_{mq}^b = \int_0^\infty \int_0^\infty \frac{\cos(d_s k_x)}{\sqrt{k_x^2+k_y^2}} \sin^2(k_x L_1/2) \frac{\sin^2(k_x L_2/2)}{k_x^2} J_0\left(k_y\frac{W_1}{2}\right) J_0\left(k_y\frac{W_2}{2}\right) dk_x\, dk_y \tag{4.34}
$$

Other remaining submatrices, such as Z_{mc}, Z_{ms}, Z_{pc} Z_{ps}, Z_{mtc}, Z_{mts}, Z_{ptc}, and Z_{pts}, involve the following two general asymptotic functions given by [91], [92]

$$
I_{mcs}^a = \int_0^\infty \int_0^\infty \frac{\cos(d_s k_x)}{\sqrt{k_x^2+k_y^2}} \frac{\sin^2(k_x L_i/2)}{k_x^2(k_x^2-k_{e_l}^2)} J_0\left(k_y\frac{W_i}{2}\right) J_0\left(k_y\frac{W_l}{2}\right) dk_x\, dk_y \tag{4.35}
$$

$$
I_{mcs}^b = \int_0^\infty \int_0^\infty \frac{\cos(d_s k_x)}{\sqrt{k_x^2+k_y^2}} \frac{\sin^2(k_x L_i/2)}{k_x^2-k_{e_l}^2} J_0\left(k_y\frac{W_i}{2}\right) J_0\left(k_y\frac{W_l}{2}\right) dk_x\, dk_y \tag{4.36}
$$

where i and l have 1 for the left signal line, and 2 for the right signal line.

From the analytical technique presented in [91], the infinite double integrals of (4.33)-(4.36) are transformed into finite one-dimensional integrals as [92]

$$I_{mq}^a = \frac{1}{\pi} \int_{-L_1-L_2}^{L_1+L_2} B(\chi - d_s) \cdot \mathcal{A}(\chi)\, d\chi \tag{4.37}$$

$$I_{mq}^b = \frac{1}{\pi} \int_{-L_1-L_2}^{L_1+L_2} B(\chi - d_s) \cdot \mathcal{B}(\chi)\, d\chi \tag{4.38}$$

$$I_{mcs}^a \approx \frac{1}{\pi} \int_{A^L}^{A^U} B(\chi - d_s) \cdot \mathcal{C}(\chi)\, d\chi - \frac{\pi}{4k_{e_l}^3}\mathcal{H} \tag{4.39}$$

$$I_{mcs}^b \approx \frac{1}{\pi} \int_{A^L}^{A^U} B(\chi - d_s) \cdot \mathcal{D}(\chi)\, d\chi - \frac{\pi}{4k_{e_l}}\mathcal{H} \tag{4.40}$$

where $B(\chi - d_s)$, $\mathcal{A}(\chi)$, $\mathcal{B}(\chi)$, $\mathcal{C}(\chi)$, $\mathcal{D}(\chi)$, $\mathcal{H}(\chi)$, A^L, and A^U are defined in [91].

The validity of the above two formulas in (4.37) and (4.38) is verified directly by letting $W_1 = W_2$ and $L_1 = L_2$. Doing this, (4.37) and (4.38) are reduced to (4.30) and (4.31).

The newly derived formulas are applied to the efficient evaluation of matrix elements in an asymmetric gap. Fig. 4.10 shows the equivalent capacitance values for a symmetric gap discontinuity with $\epsilon_r = 8.875$, $W_1 = W_2 = d = 0.508mm$, and $f=5$GHz as a function of gap spacing. For comparisons, the results obtained by the conventional SDA using the piecewise sinusoidal basis functions [93] and those of measurements [94] are included in Fig. 4.10. The conventional SDA in [93] neglects the transverse current component. And they use the upper limit $\beta^u = 400k_0$ for the evaluation of the each submatrix in (4.32). However, the proposed method using an upper limit of $\beta^u = 50k_0$, is found to be sufficiently accurate. Our results are in excellent agreement with the data obtained in [93] and seem to be in reasonably good agreement with the experimental results of [94].

For comparison of the overall computational efficiency, we calculated the average computation times between the two methods used to obtain the predicted results of

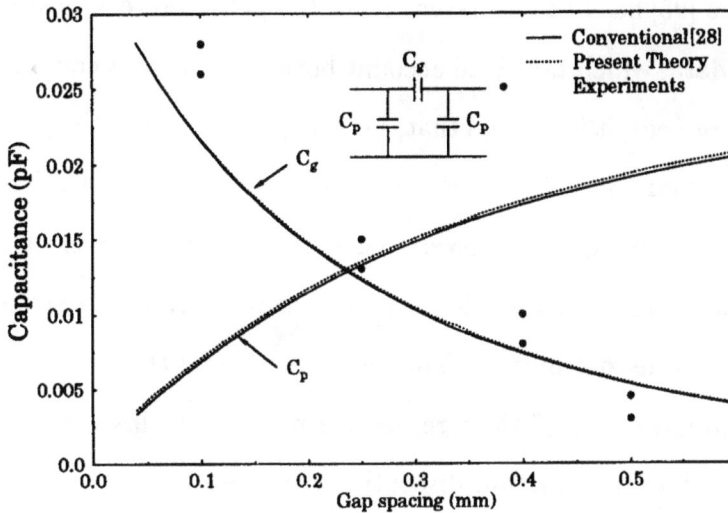

Fig. 4.10: Capacitance values of a symmetric gap with $\epsilon_r = 8.875$, $W_1 = W_2 = d = 0.508mm$

Fig. 4.10. To obtain the results of Fig. 4.10, the overall computation time of the proposed method is 17 times faster than that of the conventional method in [93].

The full-wave analysis of the gap discontinuities by considering both the transverse and longitudinal current components was reported by Jackson [48] and Alexópoulos [95]. But, on the electrically narrow strip, we can neglect the transverse current component to simplify the solution of the full-wave equation. Thus, we consider only longitudinal current component for the analysis of an asymmetric gap. Because there is a lack of data for asymmetric gaps, a symmetric gap was chosen for the initial comparison in order to validate the formulation and the computed results. Once the solution is validated for a symmetric gap, one may assume that the formulation for asymmetric gaps are also correct.

The symmetric gap on a grounded dielectric slab to be investigated has a width $W_1 = W_2 = 0.635mm$ and gap $G = 0.508mm$ with a relative permittivity $\epsilon_r = 9.9$ and a substrate thickness $d = 0.635mm$. The magnitude and phase of S_{11} and S_{21}, computed by using the proposed method, are compared with those of the full-wave solution [95], and are plotted in Figures 4.11 and 4.12. Clearly, satisfactory agreement with the results of [95] has been achieved. Since the proposed results are quite similar to the full-wave data, which take into account both an $\hat{x}$ current component and and an $\hat{z}$ current component, it is belived that, for the gap discontinuities, the assumption of using only the longitudinal current component on the narrow microstrip line is reasonable. In other words, the transverse current is negligible on electrically narrow strips. This is due to the fact that there is no mechanism to excite strong transverse currents for the gap discontinuities. Therefore we assume that this is valid even in case of the asymmetric gap, if their respective width of strips remains electrically narrow. However, for step-junction discontinuities, because of their current flow in the vicinity of the junctions, require the transverse current component in order to achieve sufficiently accurate results [95].

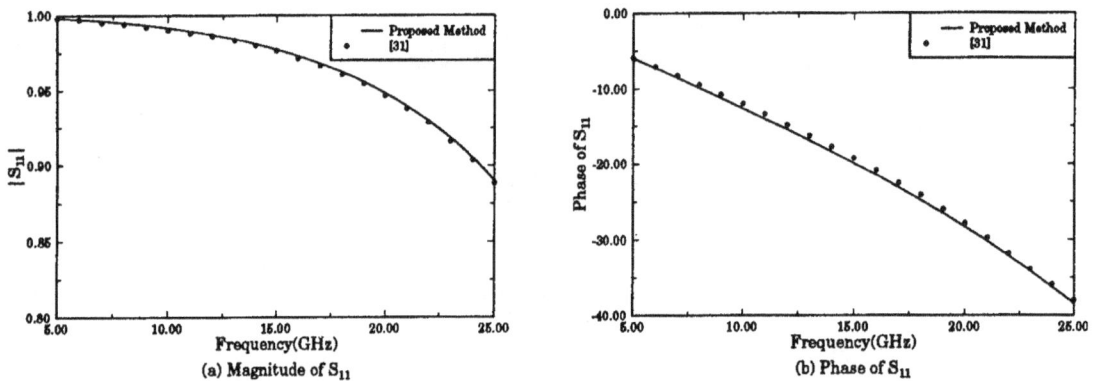

(a) Magnitude of S_{11}

(b) Phase of S_{11}

Fig. 4.11: Comparison of the S_{11} of symmetric gap($\epsilon_r = 9.9, d = 0.635mm, W_1 = W_2 = 0.635mm, G = 0.508mm$)

Next, we consider the scattering parameters of an asymmetric gap for $\epsilon_r = 6.15$, $d = 1.27mm$, $W_1 = 0.5mm$, $W_2 = 0.75mm$, and $G = 0.5mm$. In this case, the values of S_{11} are not equal to those of S_{22} due to the physical asymmetry with respect to the width of two strips. The corresponding values of S_{11}, S_{21}, and S_{22} are plotted in Fig. 4.13 as a function of frequency. Unfortunately, no computed

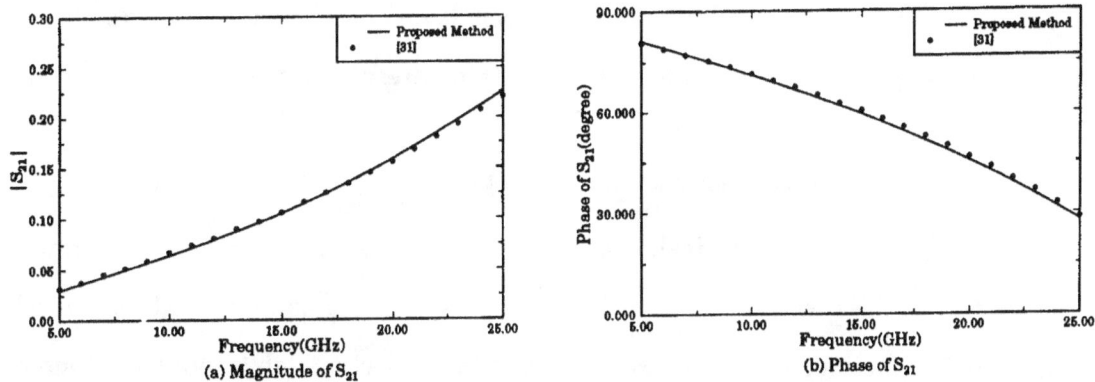

Fig. 4.12: Comparison of the S_{21} of symmetric gap($\epsilon_r = 9.9, d = 0.635mm, W_1 = W_2 = 0.635mm, G = 0.508mm$)

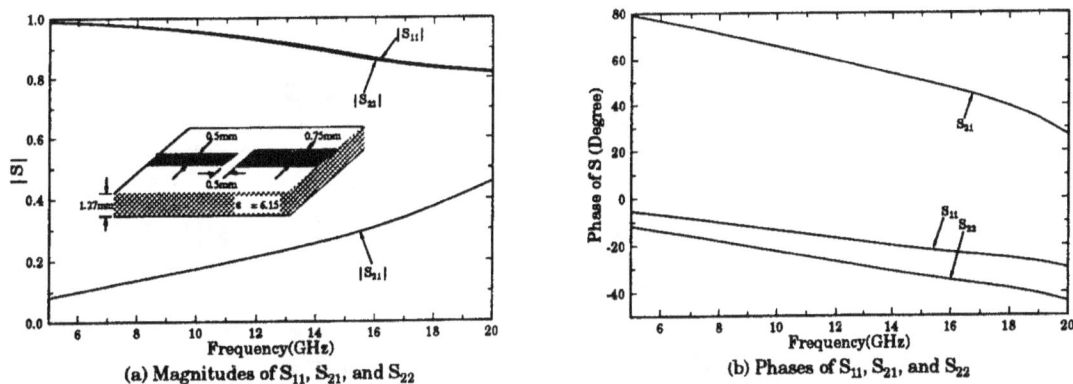

Fig. 4.13: S parameters of an asymmetric gap($\epsilon_r = 6.15$, $d = 1.27mm$, $W_1 = 0.5mm$, $W_2 = 0.75mm$, and $G = 0.5mm$).

or measured results for the asymmetric gap discontinuities could be found in the literature. Lacking such data, it is unclear to what extent the results of an asymmetric gap are accurate. It is felt that a series of experiments are needed to provide confirmation for the predicted results. However, based on the successful results of the symmetric gap, we can assume that the obtained results are reasonable and valid.

4.5.3 Scattering of rectangular microstrip patch

This part presents an analytical technique to solve the asymptotic part of the impedance matrix in the spectral domain which emploies roof-top subdomain basis functions to model surface current densities on a grounded dielectric slab. Roof-top subdomain basis functions are suitable for solving arbitrarily shaped planar geometries. However the numerical evaluation of the integrals, without an acceleration technique, leads to very time-demanding computations. The newly developed formulas presented in this work provide timely and accurate solution for the arbitrary planar circuit.

The structure investigated is a perfectly conducting rectangular patch of dimensions $W_x \times W_y$ on a grounded dielectric substrate with thickness d and dielectric constant ϵ_r shown in Fig. 4.14. Enforcing the boundary condition on the surface of a

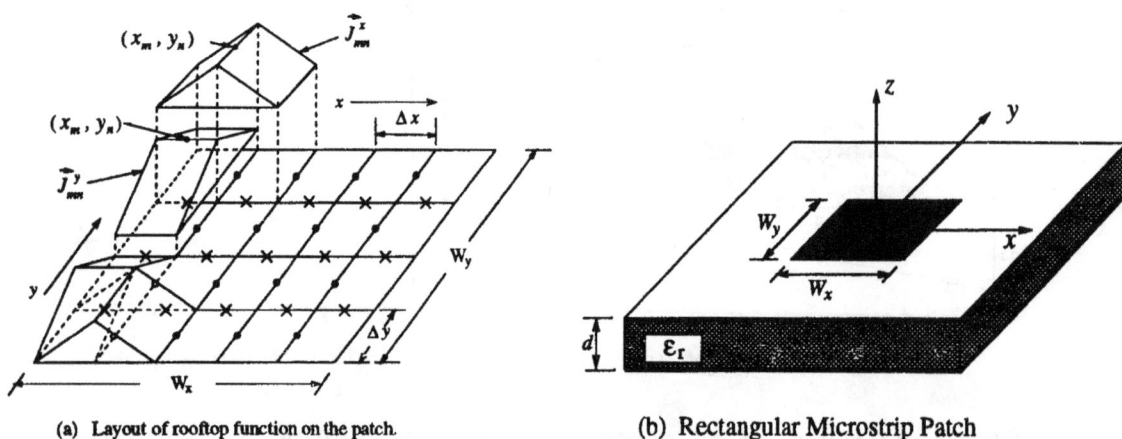

(a) Layout of rooftop function on the patch.

(b) Rectangular Microstrip Patch

Fig. 4.14: Layout of roof-top basis function on microstrip patch

perfectly conducting patch and using Galerkin's method, the electric field equation is reduced to the following matrix equations [96], [97]

$$
\begin{bmatrix} [Z^{xx}_{mnm'n'}] & [Z^{xy}_{mnm'n'}] \\ [Z^{yx}_{mnm'n'}] & [Z^{yy}_{mnm'n'}] \end{bmatrix} \begin{bmatrix} [I^{x}_{m'n'}] \\ [I^{y}_{m'n'}] \end{bmatrix} = \begin{bmatrix} [V^{x}_{mn}] \\ [V^{y}_{mn}] \end{bmatrix} \tag{4.41}
$$

where each submatrix is described in (4.14), and the unknown coefficients of $I^{x}_{m'n'}$ and $I^{y}_{m'n'}$ are given in [92], [98].

The excitation vector in the right side of (4.41) can be obtained by the inner product between the testing function and the incident field as

$$
V_{mn} = \iint\limits_{S} \vec{J}_{mn} \cdot \vec{E}^{inc} \, dx \, dy \tag{4.42}
$$

The asymptotic impedance matrix of (4.41), associated with roof-top functions and the asymptotic Green's function of (4.16)-(4.18), is written as

$$
Z^{xx^{Asy}}_{mnm'n'} = -\frac{j}{\pi^2} \frac{Z_0}{k_0} \left(\frac{8}{\Delta x} \right)^2 \left\{ -\frac{k_0^2}{2} I^{xx^a}_{mnm'n'} + \frac{1}{(\epsilon_r + 1)} I^{xx^b}_{mnm'n'} \right\} \tag{4.43}
$$

$$
Z^{xy^{Asy}}_{mnm'n'} = Z^{yx^{Asy}}_{mnm'n'} = \frac{j}{\pi^2} \frac{Z_0}{k_0} \left(\frac{64}{\Delta x \cdot \Delta y} \right) \frac{1}{\epsilon_r + 1} I^{xy}_{mnm'n'} \tag{4.44}
$$

with

$$
I^{xx^a}_{mnm'n'} = \int_0^\infty \int_0^\infty \frac{\cos(k_x x_s)}{\sqrt{k_x^2 + k_y^2}} \frac{\sin^2\left(k_y \frac{\Delta y}{2}\right)}{k_y^2} \frac{\sin^4\left(k_x \frac{\Delta x}{2}\right)}{k_x^4} \cos(k_y y_s) \, dk_x dk_y \tag{4.45}
$$

$$
I^{xx^b}_{mnm'n'} = \int_0^\infty \int_0^\infty \frac{\cos(k_x x_s)}{\sqrt{k_x^2 + k_y^2}} \frac{\sin^2\left(k_y \frac{\Delta y}{2}\right)}{k_y^2} \frac{\sin^4\left(k_x \frac{\Delta x}{2}\right)}{k_x^2} \cos(k_y y_s) \, dk_x dk_y \tag{4.46}
$$

$$
I^{xy}_{mnm'n'} = -\int_0^\infty \int_0^\infty \frac{\sin(k_x x_s)}{\sqrt{k_x^2 + k_y^2}} \frac{\sin^3\left(k_y \frac{\Delta x}{2}\right)}{k_x^2} \frac{\sin^3\left(k_y \frac{\Delta y}{2}\right)}{k_y^2} \sin(k_y y_s) \, dk_x dk_y \tag{4.47}
$$

where x_s and y_s are defined as $(x_m - x_n)$ and $(y_m - y_n)$, respectively.

With the aid of the mathematical formulas in [98], the infinite double integrals of (4.45)-(4.47) can be converted into finite one-dimensional integrals as follows:

$$
I^{xx^a}_{mnm'n'} = \frac{1}{\pi} \int_{-2\Delta x}^{2\Delta x} \mathcal{G}(\chi - x_s) \cdot \Im_a(\chi) \, d\chi \tag{4.48}
$$

$$I_{mnm'n'}^{xx^b} = \frac{1}{\pi} \int_{-2\Delta x}^{2\Delta x} \mathcal{G}(\chi - x_s) \cdot \mathfrak{S}_b(\chi) \, d\chi \tag{4.49}$$

$$I_{mnm'n'}^{xy} = \frac{1}{\pi} \int_{-\frac{3\Delta x}{2} + x_s}^{\frac{3\Delta x}{2} + x_s} \mathcal{K}(\chi) \cdot \mathcal{T}(\chi - x_s) \, d\chi \tag{4.50}$$

where $\mathcal{G}(\chi)$, $\mathcal{K}(\chi)$, and $\mathcal{T}(\chi)$ are illustrated in [92], [98].

Similar expressions are obtained for $Z_{mnm'n'}^{yy^{Asy}}$ by interchanging $\Delta x \leftrightarrow \Delta y$ and $x_s \leftrightarrow y_s$ in the (4.43), (4.48) and (4.49). If we look at the finite one-dimensional integrals of (4.48), (4.49), and (4.50) as counterparts corresponding to the double infinite integrals of (4.45), (4.46), and (4.47), we see that the interval of integration is determined only by the size of the basis function. This means that the smaller the size of the basis function, the more the interval of integration is reduced. Also if the lateral separation between any two expansion functions becomes large, the behavior of the one-dimensional integrand becomes smoother. This smooth behavior allows us to evaluate the numerical integration more accurately and timely. In addition, since the integrand of the transformed one-dimensional integral does not lead to extra calculations, it is easier to compute.

The double infinite integral in each submatrix element is carried out by the asymptotic extraction technique described in (4.15). The asymptotic impedance matrix in (4.41) is computed directly from the transformed one-dimensional integral over the finite integration region. The direct double integration of the asymptotic matrix in (4.41) is the most time consuming part of the overall computation of the matrix elements. Previously, to speed up this part, most [68], [99] have used the spatial domain method with the homogeneous Green's function which results into a four-fold integral. This technique was originally developed by Pozar [68], and requires additional computations of the same geometry in a homogeneous medium. Thus, this method is still relative time consuming. However, the calculation of this tail integral using the transformed finite one-dimensional integral has almost negligible computation time as compared to those of the first integral in (4.15).

By using the above mentioned numerical techniques, the monostatic radar cross section (RCS) of a square microstrip patch is computed as a function of frequency for the incident angles $\phi^i = 45°$, $\theta^i = 60°$ with $\hat{\theta}$ polarization, and plotted in Fig. 4.15. The structure investigated is a square patch of side $2cm$ on a grounded dielectric substrate with thickness $d = 0.07874cm$ and dielectric constant $\epsilon_r = 2.33$. As a check of the consistency and accuracy of the proposed method, our results are compared with measured data as well as those using Rao, Wilton and Glisson (RWG) subdomain basis functions which have 225 unknown coefficients [100]. Their respective results are included in Fig. 4.15. For the results of the proposed method and RWG solution, there is a good agreement with each other over a wide range of frequencies.

The convergence of the solution was investigated by varying the number of subsections. The solution of the proposed method converges relatively well by using $M = N = 12$. Also, no significant improvement in the numerical results was found by further increasing the number of M and N over the frequency range from 4GHz to 18GHz. Thus, in here, we use $M = N = 13$ which lead to $364[=M \cdot (N+1) + N \cdot (M+1)]$ unknown coefficients. To evaluate the first integral in (4.15), an upper limit $\beta^u = 50 \cdot k_0$, which gives a good convergence, is used to obtain the results in Fig. 4.15. Further increase of the upper limits does not enhance the accuracy (up to four significant figures) over the entire frequency region.

As shown in Fig. 4.15, a good agreement between the measured and the numerically computed data is observed for the lower frequencies, while for the high frequencies the agreement is less favorable. Some of difference can be attributed to physical tolerances of the experimental models which become more critical in the high-frequency region.

To illustrate the overall speed of the computation time, the CPU time between the proposed method and the conventional spectral domain approach(SDA) without acceleration, to obtain the RCS at a single frequency of 4GHz, is compared. In Table 4.3, the CPU times are given for two different number of unknowns. Using

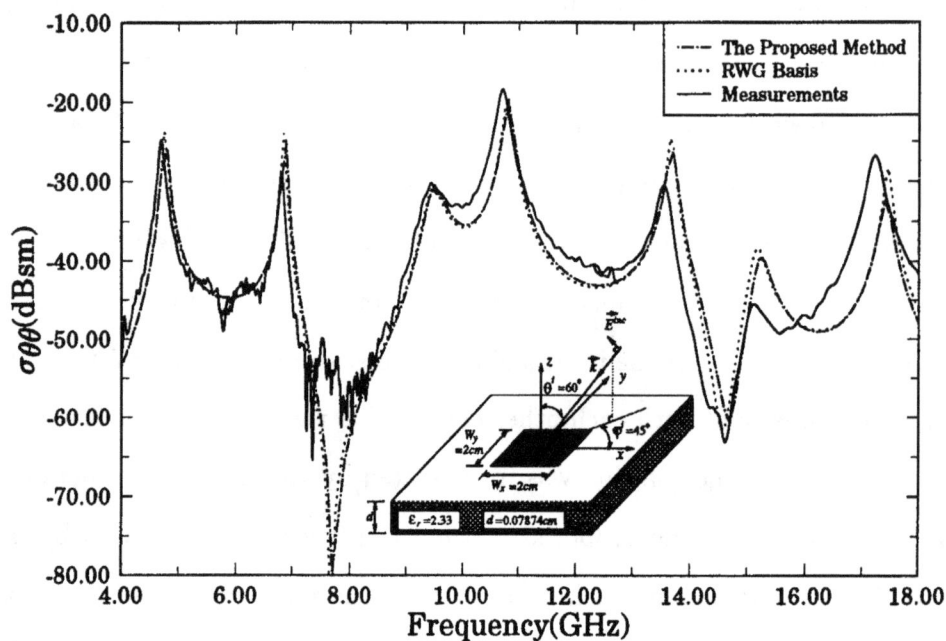

Fig. 4.15: Comparison between measurement and prediction

Table 4.3: CPU Time on a HP735/125 workstation for the Calculation of RCS (patch size $2cm \times 2cm$, $\epsilon_r = 2.33$, $d = 0.7874mm$, $\phi^i = 45°$, $\theta^i = 60°$, f=4GHz)

Number of Unknown(M=N)	SDA without acceleration[a] (seconds)	Proposed Method[b] (seconds)	Computational Efficiency $\frac{a}{b}$
180(M=9) (β^u)	553 ($300 \cdot k_0$)	68 ($50 \cdot k_0$)	8.1
364(M=13) (β^u)	2729 ($500 \cdot k_0$)	153 ($50 \cdot k_0$)	17.8

the proposed method, the chosen upper limit of $50k_0$ to evaluate the integral of the matrix elements allows the results to be accurate to four significant digits, and the overall time to obtain the RCS at 4GHz was 45 seconds with 180 unknowns. Without any acceleration, the conventional SDA does not lead to this level of accuracy until the upper limit β^u reaches $300k_0$. With this value of upper limit, approximately 553 seconds were needed using 180 unknowns. If the number of unknowns is increased to 364, small size basis functions in the conventional SDA require upper limits of $500 \cdot k_0$ to achieve a comparable accuracy, as seen in Table 4.3. The proposed method does not require that the upper limit change, depending on the size of basis functions, and it is clearly more accurate than the conventional SDA due to the elimination of the truncation error.

Although the usefulness of the newly derived formula was demonstrated through an RCS problem of microstrip patch, this approach can be easily applied to any arbitrarily shaped planar circuit.

4.6   Conclusions and future work

The main contributions of this work are the development of analytical solutions for the asymptotic matrix elements in the analysis of various geometries. A motivation for performing such procedure is to reduce the required computation time to evaluate the impedance matrix elements. Analytical techniques have been successfully used to improve the computational efficiency while retaining or even improving the accuracy, for the evaluation of the asymptotic matrix elements. This has been demonstrated successfully in a number of problems; specifically, planar transmission lines, and printed circuits including microstrip dipoles, asymmetric gap discontinuities, and scattering by a rectangular microstrip patch.

While many problems have been solved, several areas remain in need of further investigation. First it is felt that mixed grids of rectangular and triangular cells are needed to model irregular boundaries. Thus, if possible, the present work must be extended to include the subdomain basis functions such as roof-tops with rectangular and triangular support. In addition, in order to extend the capability of the formulation and corresponding computer code to analyze printed antenna, the efficient computation of the excitation matrix by closely modeling the feed region should be investigated.

Since the newly developed formulas employing the roof-top subdomain basis functions allow a large number of basis functions to be used with reasonable execution time, the contributed effort provides timely and accurate solutions than previously available acceleration techniques. These new features should be extended to develop accurate and efficient CAD tools that handle arbitrarily shaped planar circuits and analyze finite microstrip dipole arrays.

CHAPTER 5

ANALYSIS OF APERTURE ANTENNAS

5.1 Dielectric-Loaded Aperture Antennas

Aperture antennas are commonly used in the microwave frequency range. Applications range from radar tracking to missile control, navigation, satellite communications, mobile telephony, broadcast TV, and aircraft communications. Typical configurations of aperture antennas are slits, slots, waveguides, horns, reflectors, lenses, and few others. An attractive feature of these antennas is their low profile; they are usually flushmounted on the surface of large objects, such as aircraft and missiles, thereby retaining the vehicle's aerodynamic profile. The antenna opening is then covered with a thin dielectric material to provide additional protection from environmental conditions.

In this chapter, a hybrid Finite Element/Method of Moments (FEM/MoM) approach is used to analyze aperture antennas that are characterized by arbitrary shapes and material inhomogeneities. Material parameters such as permittivity and permeability are treated as full tensors, therefore, anisotropic as well as frequency dependent materials can be incorporated. Two generic types of aperture antennas considered in this study are illustrated in Fig. 5.1. The first one depicts a circular microstrip patch antenna backed by a cylindrical cavity. The patch is usually excited with a coaxial cable oriented in the vertical direction. The inner conductor is extended from the coaxial aperture to an offset point on the surface of the patch. The second configuration depicts a narrow slot backed by a rectangular cavity. A horizontal coaxial cable is used to excite the structure. Although not shown in these figures, both antennas are flushmounted on an infinite ground plane. Evaluation of the Radar Cross Section (RCS) and input impedance of these antennas was extensively performed and verified against the spectral domain method of moments, measurements performed at the ElectroMagnetic Anechoic Chamber (EMAC) at

Arizona State University, and data extracted from a variety of journal publications.

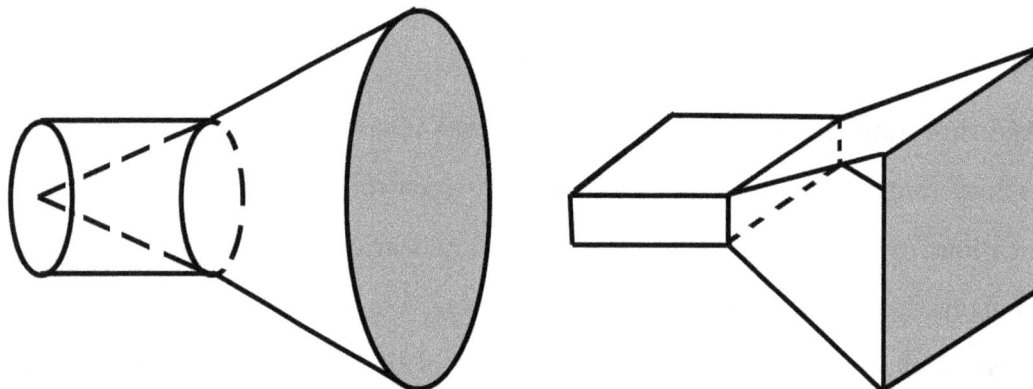

Fig. 5.1: Aperture antennas backed by a cavity. A standard 50 Ω coaxial cable is often used for excitation.

5.1.1 Radar cross section evaluation

A hybridization of the Finite Element Method (FEM) [57] and the spectral domain Method of Moments (MoM) [101], [102] is utilized in the analysis of aperture antennas mounted on an infinite ground plane. A two-dimensional view of a microstrip patch backed by an arbitrarily shaped cavity is illustrated in Fig. 5.2. The spectral domain MoM simulates field variation in the exterior of the cavity through the use of the half-space Green's function, whereas the FEM simulates field variation in the interior of the cavity using linear edge-based tetrahedral elements. The two regions are coupled through the continuity of the fields in the aperture.

The MoM formulation can be completely decoupled from the FEM formulation with the use of the physical equivalence principle after a surface magnetic current is introduced just above and below the aperture plane. The use of edge-based tetrahedral elements inside the cavity volume results in a triangular discretization of both the aperture and patch surfaces. Since one of the objectives of this study is the analysis of *arbitrary* apertures, it was decided that the most appropriate choice of basis functions for the exterior problem is the so-called Rao, Wilton and Glisson (RWG) with triangular support [103]. These are very similar to the linear edge basis functions used in the finite element analysis. The main difference between the two is that the RWG basis functions satisfy the continuity of the normal component instead of the tangential component of the field. The reason is simply because they interpolate surface magnetic current instead of electric field; the normal component of the magnetic current has to be continuous across the edge of the triangle. It is also important to emphasize here, that the implementation of the RWG basis functions is done in the spectral domain; therefore, their Fourier transform needs to be computed.

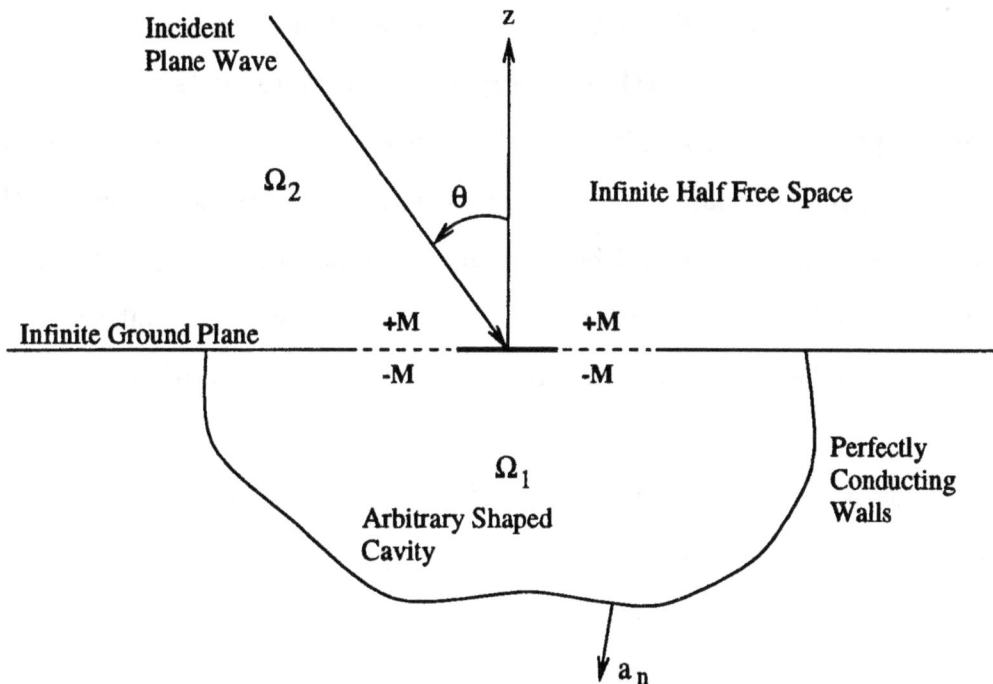

Fig. 5.2: A two dimensional cut of an arbitrary shaped cavity-backed patch antenna mounted on an infinite ground plane.

5.1.2 Validation of radar cross section analysis

A three-dimensional hybrid FEM/MoM code was written using FORTRAN 77 to compute the radar cross section of aperture antennas mounted on an infinite ground plane. The cavity is discretized using tetrahedral elements. The meshing is done with a commercial package called SDRC I-DEAS [104]. Once the mesh is completed, the boundary conditions, such as Dirichlet and Absorbing Boundary Conditions, are applied on preassigned surfaces. The mesh file is then exported in an ASCII format; specifically, a COSMIC NASTRAN format which is later read and processed by the main code. While running, the code prints out valuable information such as material definition, total number of elements, types of boundary conditions, and so on. This information can be very helpful during the debugging process. If a fundamental error occurs the code automatically terminates and prints out an error message identifying the cause. The code also writes the geometry information into a data file which can be read and displayed by GEOMVIEW; the latter is a geometry visualization package. The displayed color for surfaces depends on the applied boundary condition, thus, not only the geometry is checked but also the correctness of the imposed boundary conditions. Finally, as a post-processing step, the code is totally interfaced with PLOTMTV and TECPLOT for field intensity and current visualization at predefined surfaces.

The code is verified for various scattering problems including the radar cross section evaluation of a three-slot array backed by an air-filled rectangular cavity shown in Fig. 5.3. A frequency sweep of the array at normal angle of incidence is computed and compared with data obtained using the spectral domain method of moments. The comparison between the two data sets is illustrated in Fig. 5.4. The agreement is excellent for both polarizations. However, as the frequency increases, the discretization error might become large enough to affect the accuracy of the results.

The same geometry was reconsidered to evaluate its radar cross section versus angle at a frequency of 30 GHz. The comparison between the hybrid approach and the spectral domain MoM is depicted in Fig. 5.5. Again, an excellent agreement between the two methods is clearly shown. It is probably worth mentioning that an angle sweep simulation using an implicit numerical technique, such as the FEM or the MoM, is computationally less demanding than a frequency sweep simulation. This is because of the fact that a change on the incident angle affects only the excitation vector; thus, once the LU factorization is performed, only a back-substitution is required for subsequent angles. This is true only if an LU factorization has been used instead of an iterative solver.

The code was also verified against nonrectangular structures such as the circular patch, which is backed by a cylindrical cavity, depicted in Fig. 5.6 . The cavity itself is filled with a lossy dielectric material of $\epsilon_r = 2.2$ and $tan\delta_e = 0.0009$. The circular patch has a radius 2.5 cm whereas the cavity has a radius 3 cm; the depth of the cavity is 0.5 cm. The monostatic RCS pattern obtained using the hybrid code is compared with a pure spectral domain MoM for a wide range of frequencies. The spectral domain MoM was implemented using entire domain basis functions inside the cavity [105],[106]. As illustrated in Fig. 5.7, the two sets of data compare well with each other except near the resonant frequency. The shift in the resonant frequency is basically attributed to the insufficient discretization of the annular aperture. Further investigation of this effect proved that an increase in the number of triangular facets in the aperture tends to shift the resonant frequency of the antenna to a slightly higher value.

5.1.3   Input impedance evaluation

Although radar cross section evaluation provides a great deal of insight into the performance of an antenna, other parameters such as input impedance, radiation patterns, directivity and gain are also important and necessary for design and optimization purposes. To be able to compute all these parameters, an accurate feed

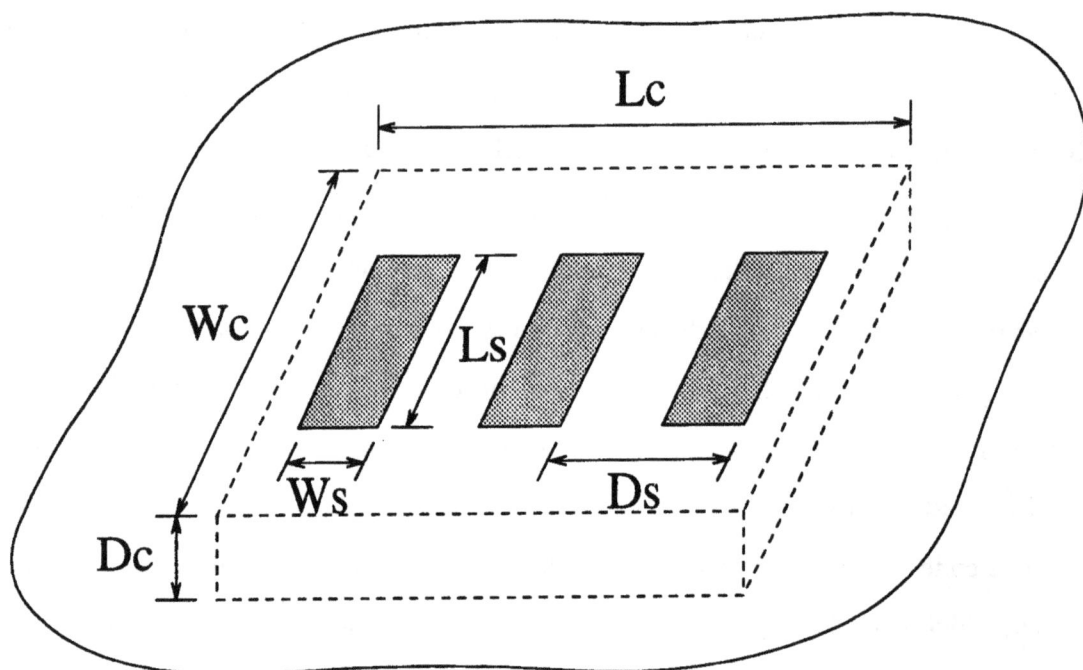

Fig. 5.3: Three-slot array backed by an air-filled rectangular cavity: $L_c = W_c = 0.75$ cm, $D_c = 0.25$ cm, $L_s = 0.5$ cm, $W_s = 0.05$ cm, $D_s = 0.25$ cm.

model had to be implemented in the hybrid FEM/MoM code. Various feed models were implemented in the past [57],[107]; however, none of them was very accurate in the context of finite elements. For example, the probe model of using the delta gap, which was first implemented in the FEM by Jin and Volakis [107], is accurate only for very thin substrates. For cases where thick substrates are present, the current along the probe is sinusoidal; therefore, the delta gap model becomes inaccurate. In addition, such an approach does not take into account the finite radius of the inner coaxial conductor.

In this section, we propose a very accurate coaxial feed model that overcomes most of the limitations found in other previously proposed feed models. The current approach is based on treating the coaxial cable as a cylindrical waveguide supporting only the dominant TEM mode. In other words, the coaxial cable is discretized using tetrahedral elements and treated as part of the finite element computational domain. A first-order absorbing boundary condition (ABC) is applied at the excitation plane

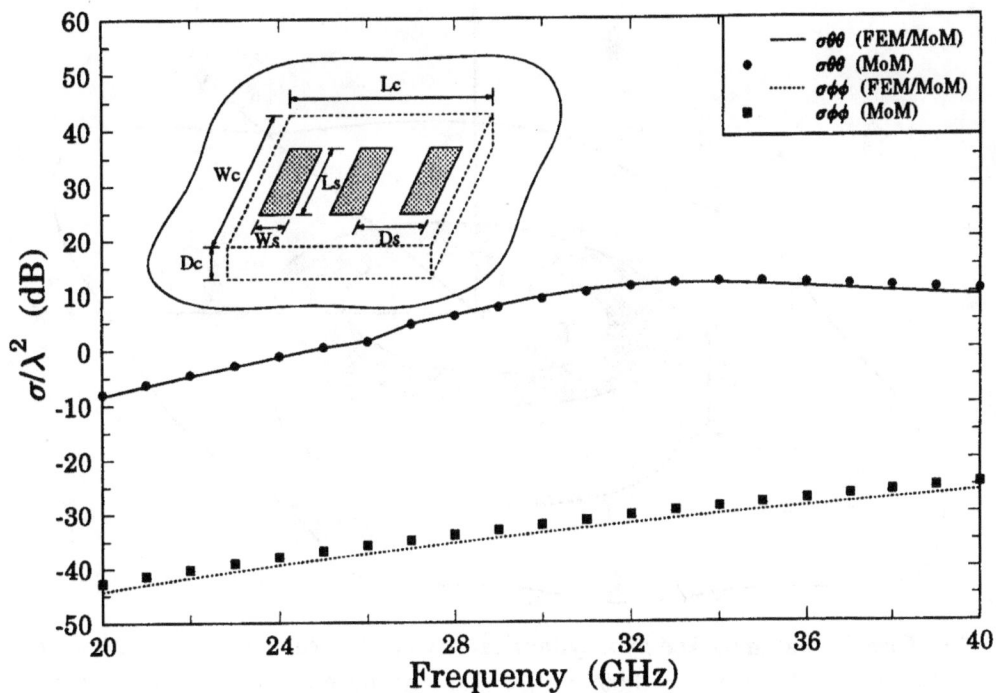

Fig. 5.4: Frequency sweep of a three-slot array backed by a rectangular cavity.

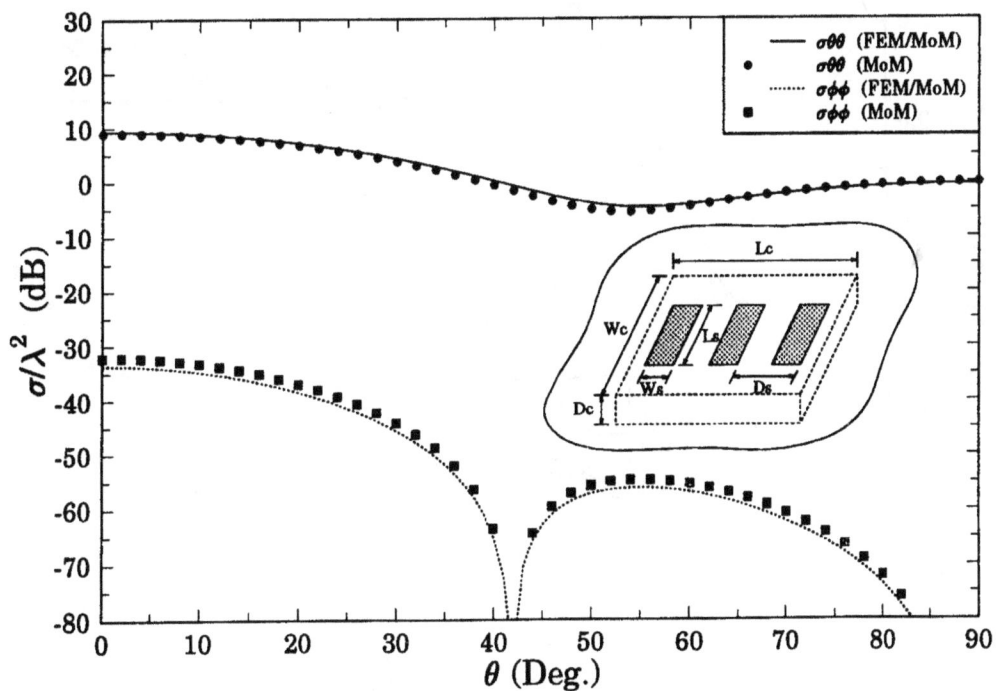

Fig. 5.5: Angle sweep of a three-slot array backed by a rectangular cavity ($f = 30$ GHz).

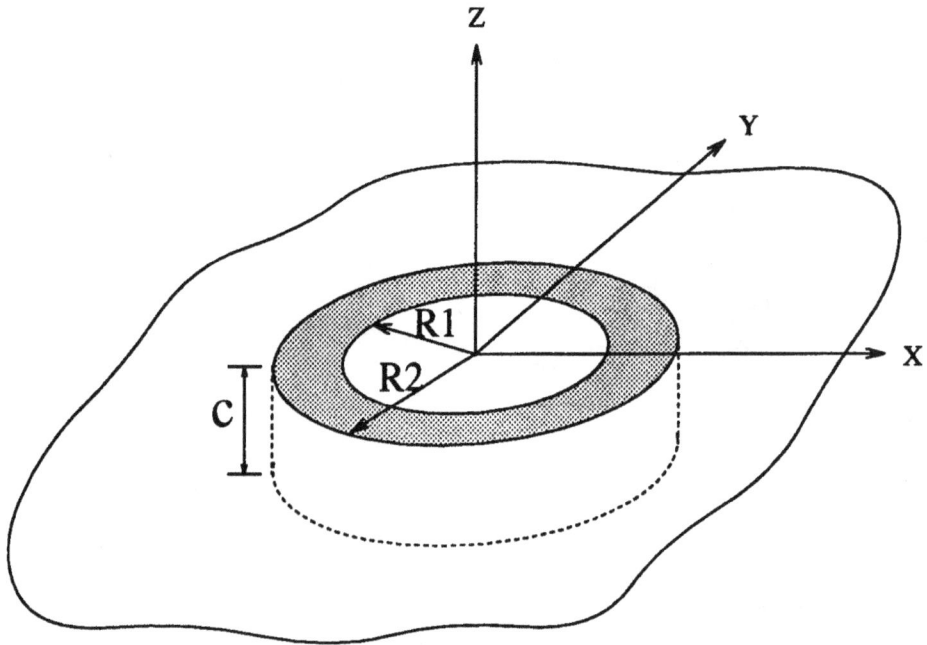

Fig. 5.6: Circular patch backed by a cylindrical cavity. The cavity is filled with a dielectric material of $\epsilon_r = 2.2$, $tan\delta_e = 0.0009$, $\mu_r = 1.0$, and $tan\delta_m = 0$. Dimensions: $R_1 = 2.5$ cm, $R_2 = 3.0$ cm, $c = 0.5$ cm.

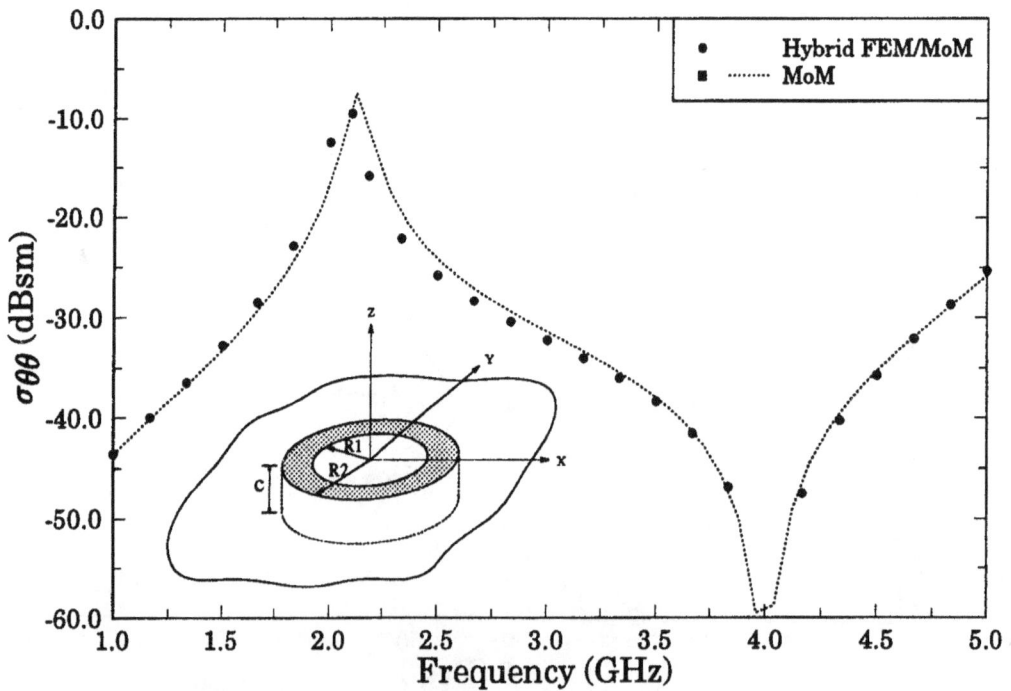

Fig. 5.7: Monostatic RCS versus frequency of a circular patch backed by a cylindrical cavity ($\theta_i = 0°$).

which usually needs to be placed at a distance approximately $\lambda/4$ away from the discontinuity.

5.1.4 Validation of input impedance analysis

The feed model that was formulated in the previous section was successfully implemented in the hybrid FEM/MoM code. The orientation of the coaxial cable can be chosen along one of the three principal directions. In order to effectively evaluate the accuracy of this feed model, it was decided that a closed empty cavity be analyzed first. A three-dimensional view of the cavity under consideration is illustrated in Fig. 5.8 whereas a detailed geometrical documentation of the problem is shown in Fig. 5.9. This geometry was chosen because of two reasons: first, the cavity is entirely closed and, thus, no radiation is released in free space; therefore, we eliminate possible numerical error, as well as experimental error, due to a strong coupling between the interior and the exterior regions of the antenna; second, the cavity is highly resonant and, thus, the accuracy of the proposed feed model is tested under the worst possible scenario. The measurements for this cavity were performed on an HP8510 network analyzer at the Arizona State University ElectroMagnetic Anechoic Chamber (EMAC). A comparison between the real and imaginary parts of the corresponding reflection coefficient, for a frequency band of 6 GHz, is illustrated in Figures 5.10 and 5.11. Although the coaxial cable was modeled only 1 cm long, the comparison between the FEM and the measurements shows an excellent agreement. During a numerical simulation, one should always make sure that the length of the coaxial cable is chosen to be long enough so that higher-order modes, which are evanescent below 35 GHz, decay before they reach the excitation plane. If the length of the coaxial cable is not large enough, these higher-order fields will reflect back toward the aperture, due to the presence of the excitation plane, thereby perturbing the TEM-like fields. Note that the absorbing boundary condition at the excitation plane is developed based on a TEM-like propagating wave; thus, it can effective in absorbing only a TEM wave. A longer coaxial cable, provided that the mesh density remains the same, results in more accurate predictions; however, the number of unknown grows significantly.

The same geometry shown in Fig. 5.8 is reexamined with one of the cavity plates completely removed so that the antenna is mounted on an infinite ground plane. Specifically, the face located at the $z = 0$ plane, which is the farthest surface away from the coaxial probe, is the one that is removed. Thus, the FEM is used to model the fields inside the cavity and the coaxial cable, whereas the spectral domain MoM is used to model the fields in the exterior region of the cavity. The input impedance of this cavity-backed slot antenna is computed within a wide frequency band. The measurements were performed using the HP8510 network analyzer at Arizona State University. As far as the experiment is concerned, the aperture antenna was mounted on a finite ground plane of dimensions $24'' \times 24''$; the sharp edges were covered with absorbing material in order to reduce diffractions. In addition, the aperture was rotated at an angle with respect to the principal axis and offset relative to the center of the ground plane; therefore, the diffractions are directed away from the antenna. The input impedance comparison between the numerical predictions and the measurements is illustrated in Figures 5.12 and 5.13. Two different simulation cases were considered: one with a coaxial cable of length $L_c = 8$ cm, and another with a coaxial cable of length $L_c = 8$ cm. Both cases show an excellent agreement with the measurements. The slight shift discrepancy is most likely attributed to a reference plane mismatch in the measurements. However, if the coaxial cable is modeled much shorter, e.g. $L_c = 1$ cm, significant discrepancies start appearing at the higher frequencies. Such observation is illustrated in Figures 5.14 and 5.15. As it was said before, this discrepancy between predictions and measurements, when the coaxial cable is electrically short, is a result of higher-order modes that reach the excitation plane.

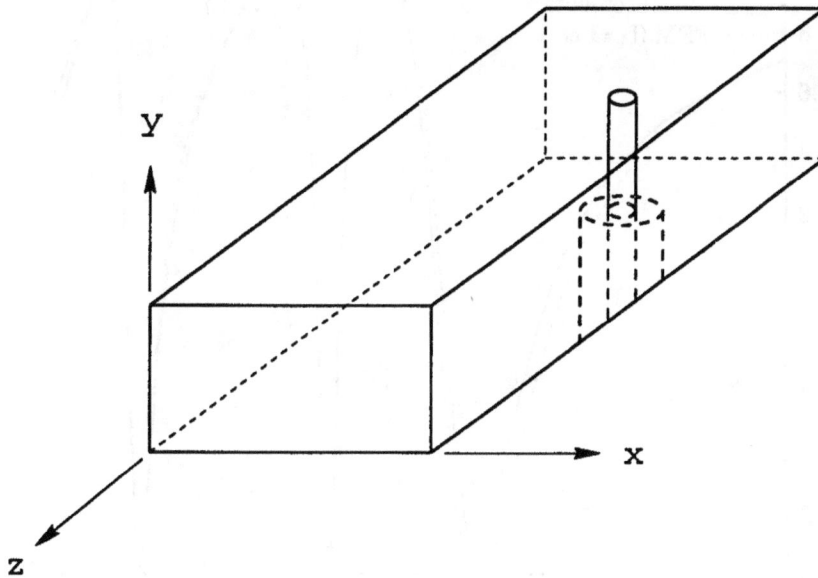

Fig. 5.8: A three-dimensional view of an air-filled rectangular cavity fed with a 50Ω coaxial cable oriented in the y-direction.

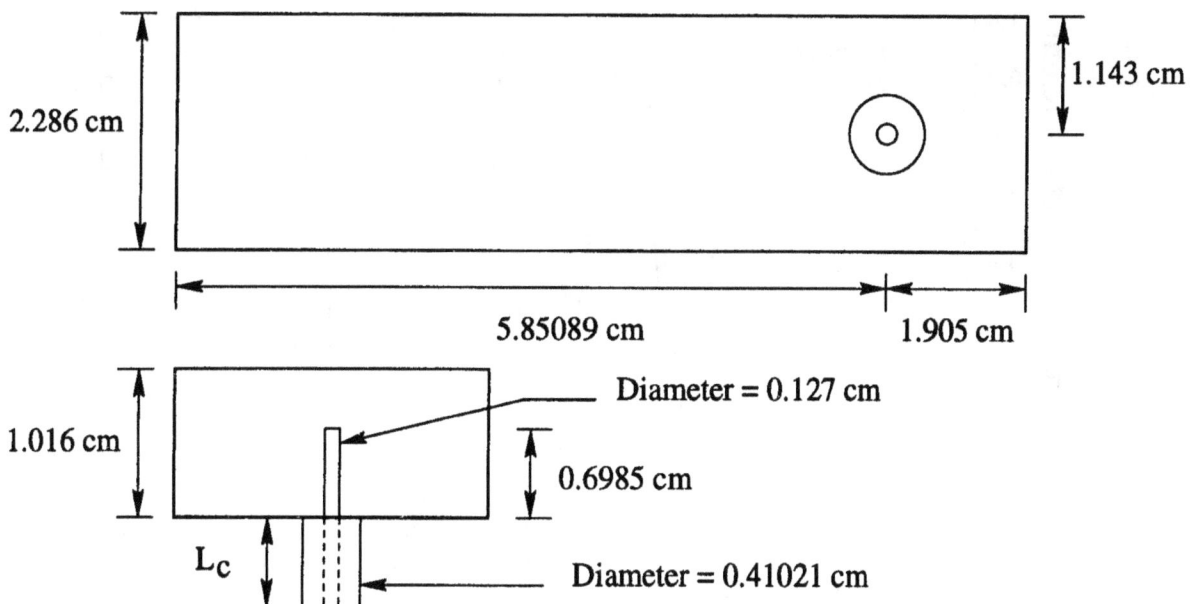

Fig. 5.9: A two-dimensional view of an air-filled rectangular cavity fed with a 50Ω coaxial cable oriented in the y-direction.

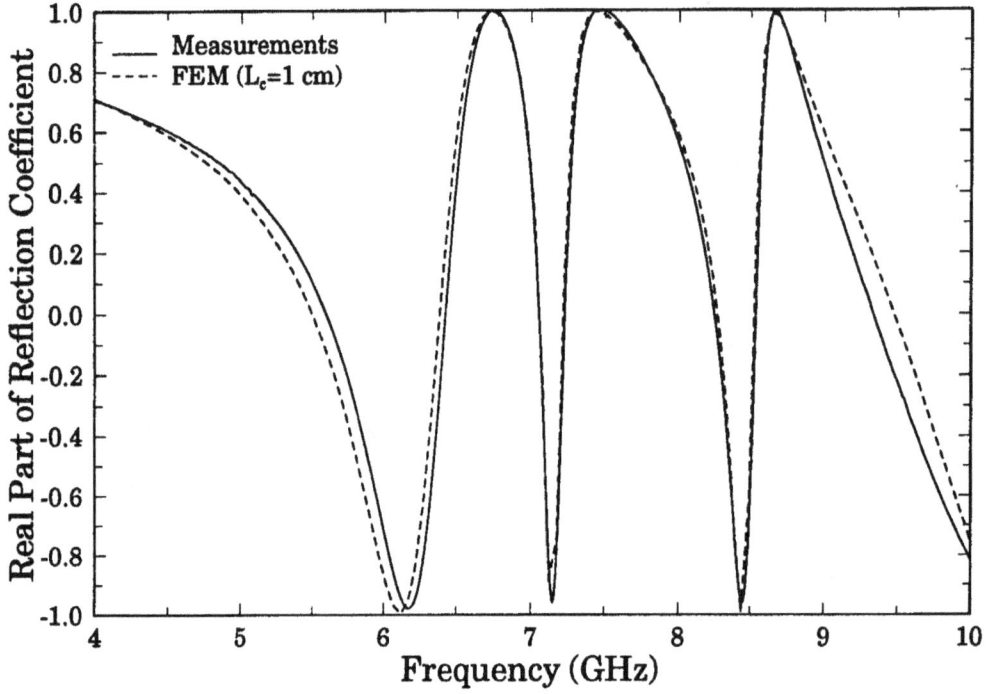

Fig. 5.10: Real part of the reflection coefficient of a closed empty rectangular cavity fed with a 50Ω coaxial cable.

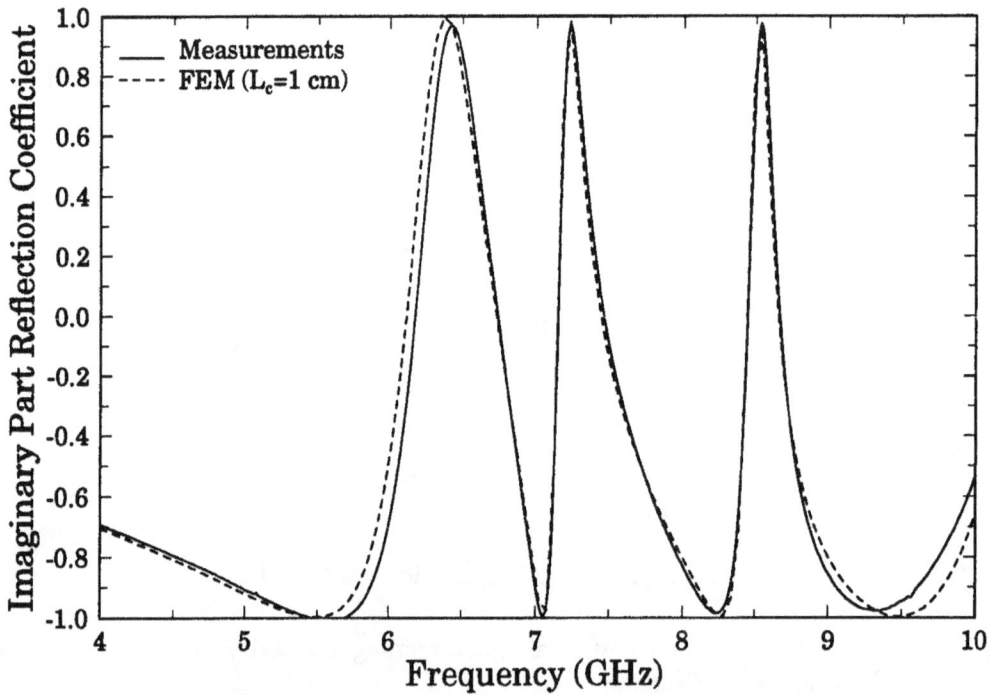

Fig. 5.11: Imaginary part of the reflection coefficient of a closed empty rectangular cavity fed with a 50Ω coaxial cable.

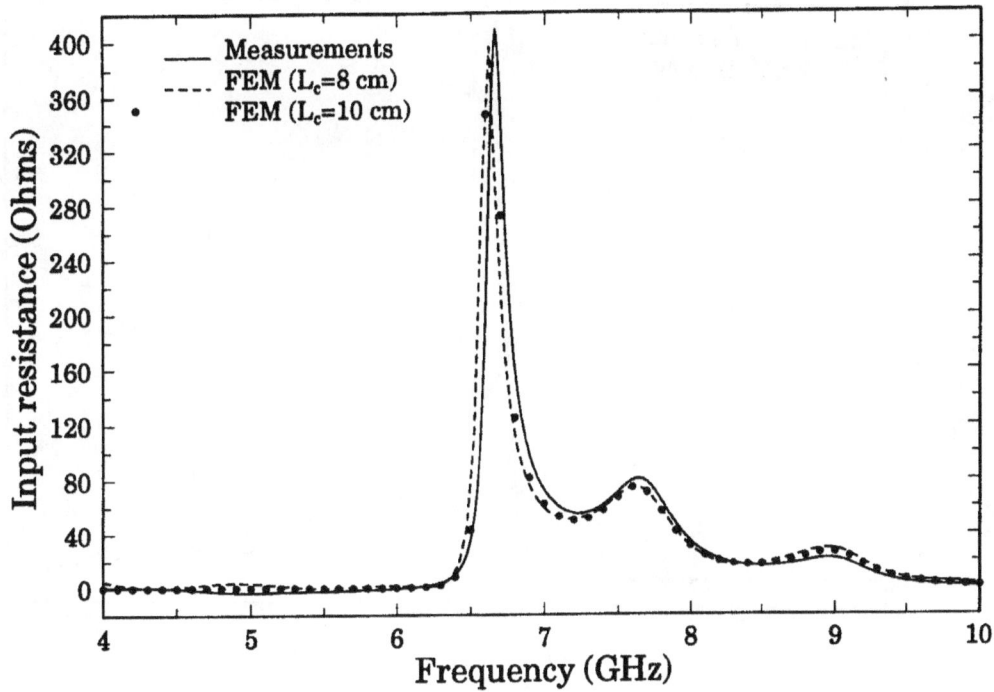

Fig. 5.12: Input resistance of an air-filled cavity-backed slot antenna fed from the side with a 50Ω coaxial cable.

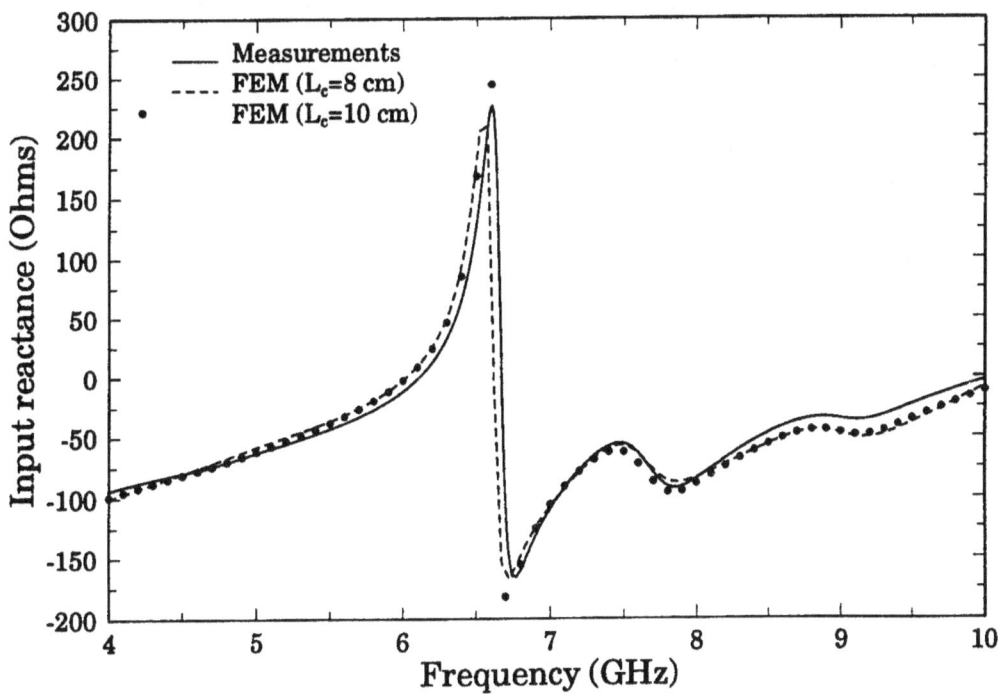

Fig. 5.13: Input reactance of an air-filled cavity-backed slot antenna fed from the side with a 50Ω coaxial cable.

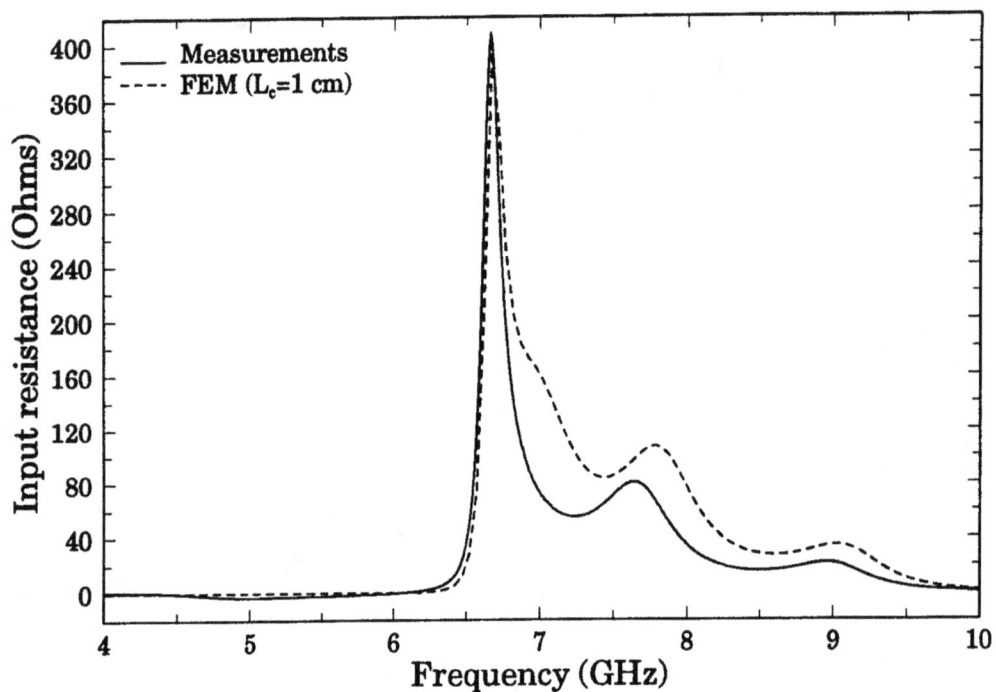

Fig. 5.14: The effect on the input resistance after shortening the coaxial cable from 8 cm to 1 cm.

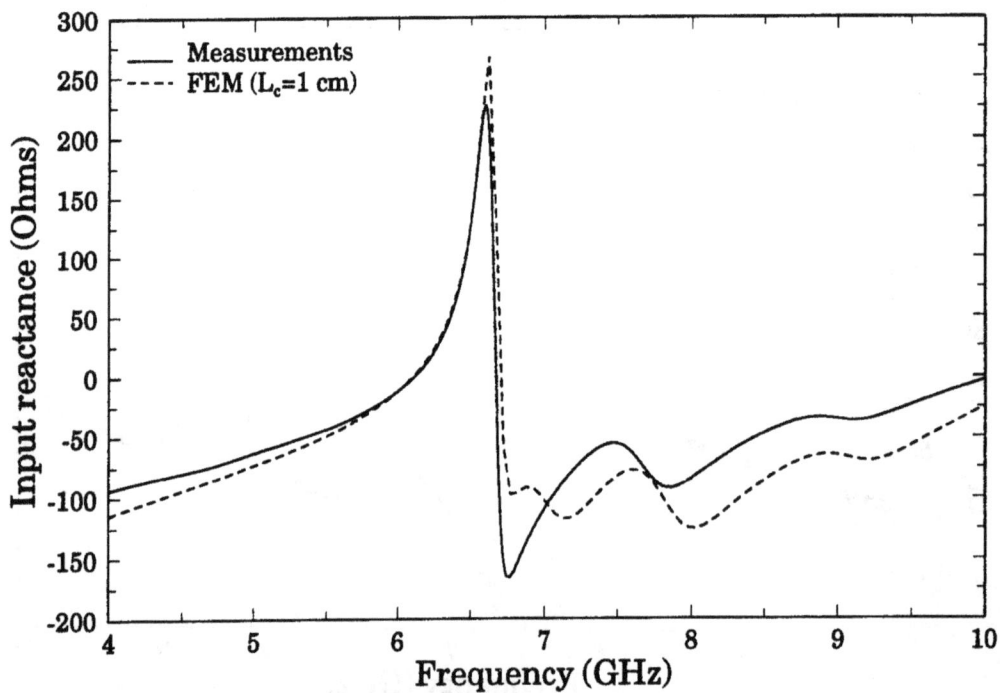

Fig. 5.15: The effect on the input reactance after shortening the coaxial cable from 8 cm to 1 cm.

5.2 Ferrite-Loaded Aperture Antennas

Ferrites have been used for many years in microwave and millimeter-wave devices such as circulators, isolators, and phase shifters [108]. The integration of ferrite technology in monolithic microwave integrated circuits (MMICs) has many advantages including design diversity and versatility. An important property of biased ferrites is that material parameters are nonreciprocal and electronically tunable. For example, in ferrite-loaded waveguide structures the forward and backward waves have different propagation characteristics. This effect can be utilized to design devices such as isolators and circulators.

Recently, great interest has been shown in using magnetized ferrites as substrates or superstrates for microstrip patch antennas [109],[110]. Gyromagnetic materials have also been used for cavity-backed aperture antennas [111]. It has been demonstrated both experimentally and analytically [112]-[114] that antenna characteristics such as resonant frequency, radar cross section, gain and axial ratio are strongly dependent on the direction and strength of the biased field. Also, it has been shown that ferrite materials might cause additional resonances to appear at the lower range of frequencies; these are attributed mainly to the existence of magnetostatic modes inside the ferrite layer [111].

To better understand wave propagation inside ferrite materials, consider a linearly polarized plane wave at normal incidence on a transversely biased ferrite slab. Two different type of waves are excited inside the gyromagnetic medium; these are known as the ordinary and the extraordinary waves [115], [114]. The ordinary wave is simply the same as the plane wave propagating inside a dielectric material. This type of wave is totally unaffected by the magnetization of the ferrite. On the other hand, the extraordinary wave is propagating along the direction of the biased magnetic field, thereby affecting its propagation characteristics. The propagation constant of the extraordinary wave is given by [115]

$$\beta_e = \omega\sqrt{\epsilon\mu_{eff}} \tag{5.1}$$

with

$$\mu_{eff} = \frac{\mu^2 - \kappa^2}{\mu} \tag{5.2}$$

$$\mu = \mu_o\left(1 + \frac{\omega_o\omega_m}{\omega_o^2 - \omega^2}\right) \tag{5.3}$$

$$\kappa = \mu_o\left(\frac{\omega\omega_m}{\omega_o^2 - \omega^2}\right) \tag{5.4}$$

where $\omega_o = \mu_o\gamma(H_o + j\frac{\Delta H}{2})$ and $\omega_m = \mu_o\gamma M_s$. In these expressions, the externally applied magnetic field is denoted as H_o, the linewidth of the ferrite material as ΔH, and the saturization magnetization as $4\pi M_s$. The ordinary and extraordinary waves have field components that are perpendicular with each other. In other words, if an incident wave is polarized in the direction of magnetization the propagating wave inside the transversely biased ferrite will have all properties of the extraordinary wave. On the other hand, if the incident wave is polarized in the direction perpendicular to the direction of magnetization the propagating wave will behave like an ordinary wave. The two type of waves are completely decoupled only for the case of normal incidence. If an incident wave impinges the ferrite material at an angle, then the ordinary and extraordinary waves are coupled. Also, from (5.2) it is apparent that the effective permeability μ_{eff} may become negative for certain values of ω, ω_o and ω_m. In such a case, the propagation constant also becomes negative, therefore, the wave attenuates rapidly (evanescent wave) as it penetrates the ferrite slab. This phenomenon is usually referred to as the cut-off state of the ferrite material. An incident wave polarized along the direction of magnetization will be totally reflected if μ_{eff} becomes negative. The frequency range where μ_{eff} is negative is given by

$$\sqrt{\omega_o(\omega_o + \omega_m)} \leq \omega \leq \omega_o + \omega_m. \tag{5.5}$$

Besides transverse-plane magnetization, the ferrite slab may be otherwise magnetized along the direction of propagation. The main observation is that the forward traveling wave is propagating with a different propagation constant than the backward traveling wave. The corresponding expressions are given by

$$\beta_+ = \omega\sqrt{\epsilon(\mu + \kappa)} \tag{5.6}$$

$$\beta_- = \omega\sqrt{\epsilon(\mu - \kappa)}. \tag{5.7}$$

Not only the propagation constant is different for the forward and backward waves, but also the attenuation constant, assuming that the ferrite material exhibits some type of loss. Also, when one wave exhibits a right-hand circular polarization, the other wave always exhibits a left-hand circular polarization. The superposition of the two propagating waves however, still represents a linear polarization.

The permeability of a magnetized ferrite is numerically modeled using a tensor notation. Depending on the direction of the biased field, the structure of the tensor is different. For example, when a ferrite material is magnetized in the $z-$direction, the complex permeability tensor is expressed as

$$[\mu] = \begin{bmatrix} \mu & -j\kappa & 0 \\ j\kappa & \mu & 0 \\ 0 & 0 & \mu_\circ \end{bmatrix} \tag{5.8}$$

where μ and κ are explicitly given in (5.3) and (5.4). In case the direction of magnetization is along the $x-$ or $y-$axis, the ferrite permeability tensor has to be rotated by 90 degrees.

As mentioned at the beginning of this discussion, full-wave numerical techniques, such as the spectral domain method of moments, have been extensively used to analyze microstrip patch antennas on ferrite substrates [109]-[114]. All this previous work was primarily concentrated in predicting and analyzing the radar cross section of these antennas. The recent work by Kokotoff [111] is probably the only published document that investigated in detail the use of magnetized ferrites in designing, fabricating and testing a feasible antenna. Nevertheless, some of the major conclusions of previous research studies related to ferrite-loaded antennas include tunability, circular polarization, beam steering, RCS control, surface wave reduction and gain enhancement. Ferrite materials in antenna technology require further attention, especially in cases where the antenna is a radiator instead of a scatterer. An extensive investigation of ferrite loaded antennas needs to be actively pursued, such as altering the direction of magnetization, the strength of the external biased field, the saturization magnetization, *etc.*. Observations on the variation of input impedance, radiation patterns, directivity, and gain in the presence of ferrite materials should also be carefully analyzed for the development and design of more practical and cost effective antennas which would operate at low frequencies (VHF/UHF) and still retain most of their high frequency characteristics such as high gain and broad bandwidth. The ability to easily tune ferrite-loaded microstrip patch and aperture antennas, in addition to numerous other advantages, provides an additional advantage for their implementation in commercial and military applications.

In this study, the finite element method hybridized with the method of moments is used to analyze cavity-backed aperture antennas loaded with layers of ferrite materials. Unlike pure method of moments, the current analysis allows the use of any direction of magnetization. In addition, arbitrary shapes of cavities and apertures may be considered. The following section is solely focussed on radar cross section analysis of ferrite-loaded antennas.

5.2.1 Radar cross section of ferrite-loaded aperture antennas

A three-dimensional finite element code, which was fully hybridized with the spectral and spatial domain method of moments, was used to evaluate the radar cross section of cavity-backed aperture antennas loaded with ferrite materials. The code was validated by predicting the RCS of the multi-ferrite-layer cavity-backed slot antenna shown in Fig. 5.16. This antenna was originally designed, built and tested by Kokotoff [111] using both experiments and full-wave numerical techniques. The cavity volume is partitioned horizontally into five sections. Each section is filled with either dielectric or ferrite material. The numbering of the material layers starts in ascending order from top to bottom. Material parameters and dimensions of this geometry are tabulated in Table 5.1. The monostatic RCS of this antenna was calculated versus frequency at normal incidence. The ferrite material was magnetized in the $y-$direction with an external DC magnetic field of $H_0 = 400$ Oe. The predicted results ($\sigma_{\phi\phi}$ polarization) using the hybrid FEM/MoM code are compared with data extracted from [111]. As depicted in Fig. 5.17, the two data sets are in excellent agreement. The MoM data shown in this figure are plotted only up to 850 MHz; the reason is because the method becomes unstable at higher frequencies, primarily due to instabilities in the Green's function. This can be thought of as another advantage of using the FEM to treat the ferrite-loaded cavity instead of the MoM approach.

It is probably important to mention here that the frequency range in which the extraordinary wave starts attenuating is dependent on the actual ferrite parameters and the strength of the biased field. This frequency range can be precisely estimated using the formula in (5.5). Concerning the ferrite-loaded antenna shown in Fig. 5.16, the extraordinary wave will start decaying approximately between 2.0 GHz and 3.4 GHz. Within this frequency band, a significant drop of the radar cross section may be observed depending on the polarization of the incident field. In such a case, it is possible that additional resonances might appear in the lower or the upper range of frequencies. This property of ferrites have been utilized in the past to design switchable microstrip patch antennas.

The finite element predictions shown in Fig. 5.17 were obtained by running the code on a 370 IMB RISC 6000 workstation. The three-dimensional mesh was created using a commercial package called SDRC I-DEAS. The total number of tetrahedral elements was 7356, whereas the number of unknowns was 8006. The remaining

Table 5.1: Antenna dimensions and material specifications.

Variable	Meaning	Dimensions
a	cavity width	5.080 cm
b	cavity length	5.080 cm
c	cavity depth	5.080 cm
τ_1	layer thickness	0.726 cm
μ_{r1}	relative permeability	1.00
ϵ_{r1}	relative permittivity	2.20
τ_2	layer thickness	1.790 cm
μ_r	relative permeability	1.0
$4\pi M_s$	saturation magnetization	800 G
ΔH	resonant linewidth	5 Oe
H_o	assumed internal DC bias field	400 Oe
ϵ_{r2}	relative permittivity	13.90
τ_3	layer thickness	0.737 cm
μ_{r3}	relative permeability	1.00
ϵ_{r3}	relative permittivity	2.20
τ_4	layer thickness	0.762 cm
μ_r	relative permeability	1.0
$4\pi M_s$	saturation magnetization	800 G
ΔH	resonant linewidth	5 Oe
H_o	assumed internal DC bias field	400 Oe
ϵ_{r4}	relative permittivity	13.90
τ_5	layer thickness	1.065 cm
μ_{r5}	relative permeability	1.00
ϵ_{r5}	relative permittivity	1.00

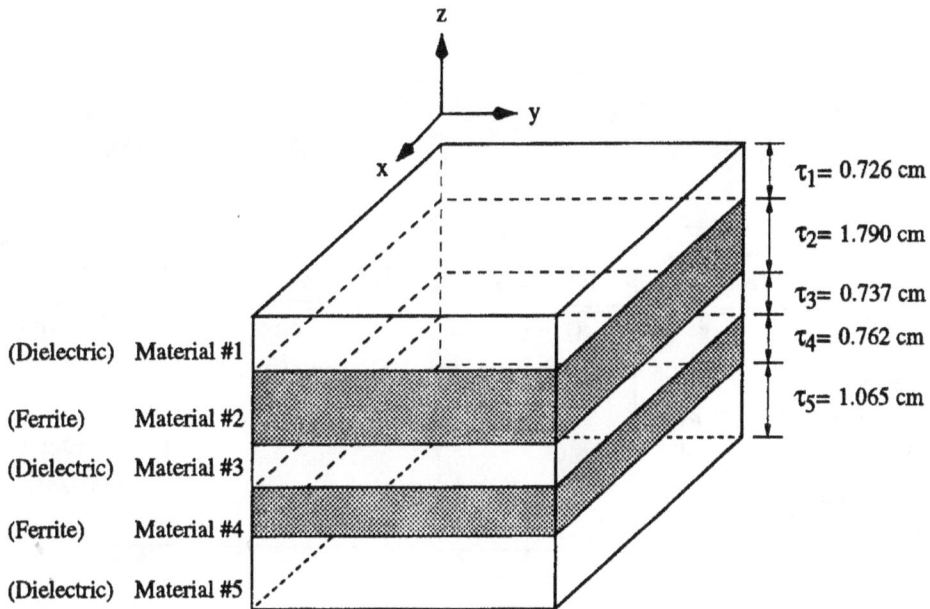

Fig. 5.16: Multi-layer, ferrite-loaded cavity-backed slot antenna.

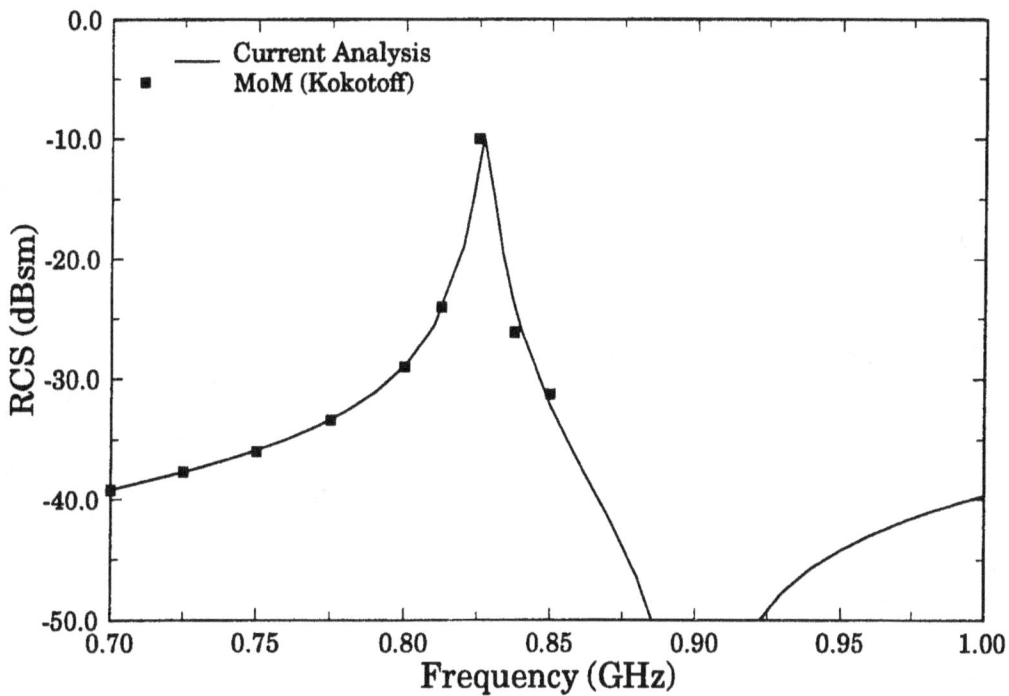

Fig. 5.17: Data comparison between FEM and MoM for the multi-layer, ferrite-loaded aperture antenna. Ferrite parameters: $H_o = 400$ Oe, $\Delta H = 5$ Oe, $4\pi M_s = 800$ G, $\epsilon_r = 13.9$.

Table 5.2: Computational statistics of the hybrid FEM/MoM code.

Problem Parameters	Total CPU Time	External Integration Time	CGS Solver Time
Elements = 7356 FEM Unknowns = 8006 MoM Unknowns = 209 Solution Tol. = 1.0e-5 Evaluation pnts. = 61 IBM 370 Risc 6000	31 hours (30 min/pnt.)	25 hours (25 min/pnt.)	3 hours (3 min/pnt) Including both polarizations

Table 5.3: Computational statistics of the hybrid FEM/MoM code after a frequency interpolation of the admittance matrix is introduced.

Problem Parameters	Total CPU Time	External Integration Time	CGS Solver Time
Elements = 7356 FEM Unknowns = 8006 MoM Unknowns = 209 Solution Tol. = 1.0e-5 Evaluation pnts. = 61 Interpolation based on 3 frequency points IBM 370 Risc 6000	5:12 hours:min (5 min/pnt.)	1:15 hours:min	3 hours (3 min/pnt) Including both polarizations

setup parameters for the problem are given in Table 5.2. The monostatic radar cross section of the antenna was evaluated at 61 frequency points. The total run time was 31 hours which is equivalent to approximately 30 minutes per frequency point. However, a closer look at the computational statistics indicates that most of the CPU time (25 minutes per point) was spent to fill in the MoM admittance matrix, which represents the exterior part of the problem. It was then decided that a linear interpolation of the admittance matrix be used across the frequency spectrum. The previous simulation was repeated, but now the admittance matrix is evaluated only at three frequency points; at in-between points, the entries of this matrix are linearly interpolated. The remaining settings of the problem as well as the corresponding computational statistics are illustrated in Table 5.3. Specifically, the hybrid code now takes only a total of 5 hours and 12 minutes to compute the monostatic RCS of the antenna for 61 frequency points. In other words, using linear interpolation for the admittance matrix, the hybrid code requires on the average only 5 minutes per point instead of 30 minutes per point observed when no interpolation is applied. In terms of accuracy, the results in both cases are identical.

The ability to effectively tune the ferrite-loaded cavity-backed slot antenna, shown in Fig. 5.16, was tested by altering the externally biased magnetic field H_o. The strength of the magnetic field was constantly increased from 400 Oe to 700 Oe. As illustrated in Figs. 5.18 and 5.19, the resonant frequency of the antenna moves to a higher frequency as H_o increases. The reason for this frequency shift is related to the inherent properties of the ferrite material which are set by the structure of the permeability tensor. In order for someone to grasp a better understanding of the wave behavior inside a ferrite, the effective permeability, μ_{eff}, of the extraordinary wave as it propagates inside a homogeneous ferrite medium of $\epsilon_r = 13.0$, $\Delta H = 0$ Oe, and $4\pi M_s = 800$ G is plotted as the external magnetic field is varied from 0 to 800 Oe. The corresponding graph is depicted in Fig. 5.20. When $H_o = 0$ Oe, according to formula (5.5), μ_{eff} is negative in the frequency range between dc and 2.24 GHz. In other words, the extraordinary wave attenuates in this frequency band as it propagates inside the ferrite. As the frequency increases, the value of μ_{eff} asymptotically approaches 1. On the other hand, as the external magnetic field increases, ω_0 which determines the lower end of the *attenuation band* becomes nonzero. The most interesting behavior of the ferrite tensor however, according to Fig. 5.20, is that μ_{eff} is positive for all frequencies outside the *attenuation band* and negative for all frequencies inside the *attenuation band*. Also, the transition between positive and negative values of μ_{eff} is very abrupt due to the presence of a frequency pole. Near this transition region, the wavelength of the extraordinary wave can change quite rapidly. Thus, depending on the antenna configuration, additional resonances might appear in that frequency region. However, inside the *attenuation band*, the antenna response is usually very low even though the ferrite material might be totally lossless. This phenomenon is sometimes referred to as the *cut-off state* of a ferrite-loaded antenna.

Referring once again to Fig. 5.20, it is interesting to see that an increase in the external magnetic field results in a significant shift of the *attenuation band* and the transition region to a higher frequency. Thus, if the antenna resonates at a frequency in the vicinity of the transition region, which is usually the case, a variation in the external field would effectively tune the antenna within a certain frequency band. Note that such a resonance appears due to the presence of the extraordinary wave and not the ordinary wave. The extraordinary wave, as was previously mentioned, strongly depends on the polarization of the propagating wave. In case that the polarization does not excite a strong extraordinary wave, the corresponding resonant peak will be significantly low. In other words, understanding the behavior of the extraordinary wave inside a ferromagnetic medium does not always guarantee accurate prediction of strong magnetostatic modes.

Besides increasing the externally biased magnetic field, the saturation magnetization $4\pi M_s$ was also increased while observing the variation in the effective

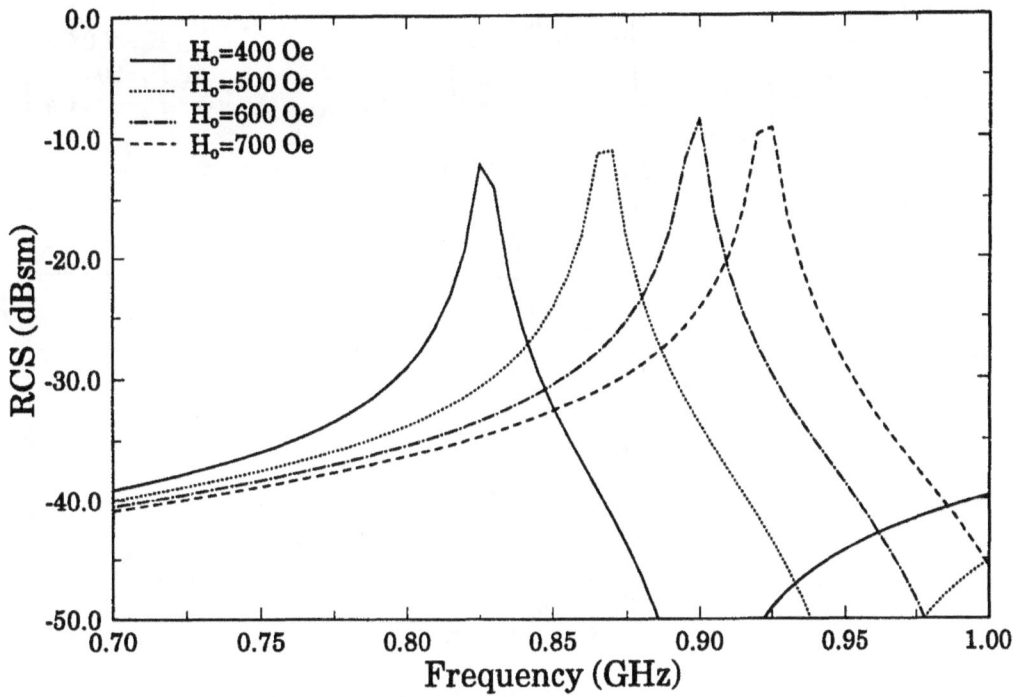

Fig. 5.18: Frequency tuning of the antenna by varying the external magnetic field ($\sigma_{\phi\phi}$).

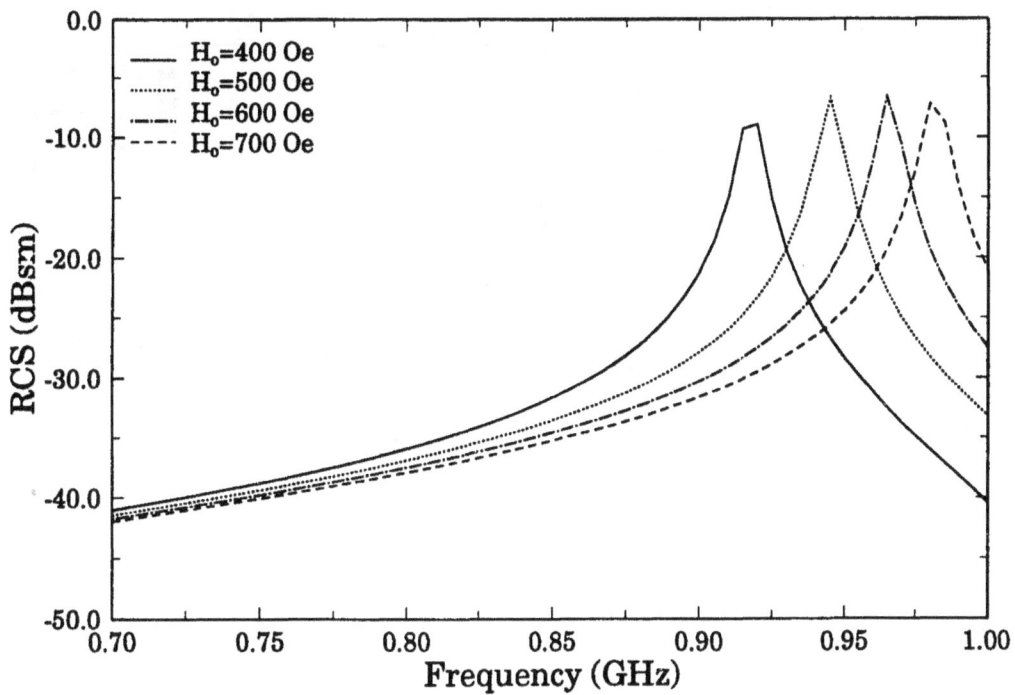

Fig. 5.19: Frequency tuning of the antenna by varying the external magnetic field ($\sigma_{\theta\theta}$).

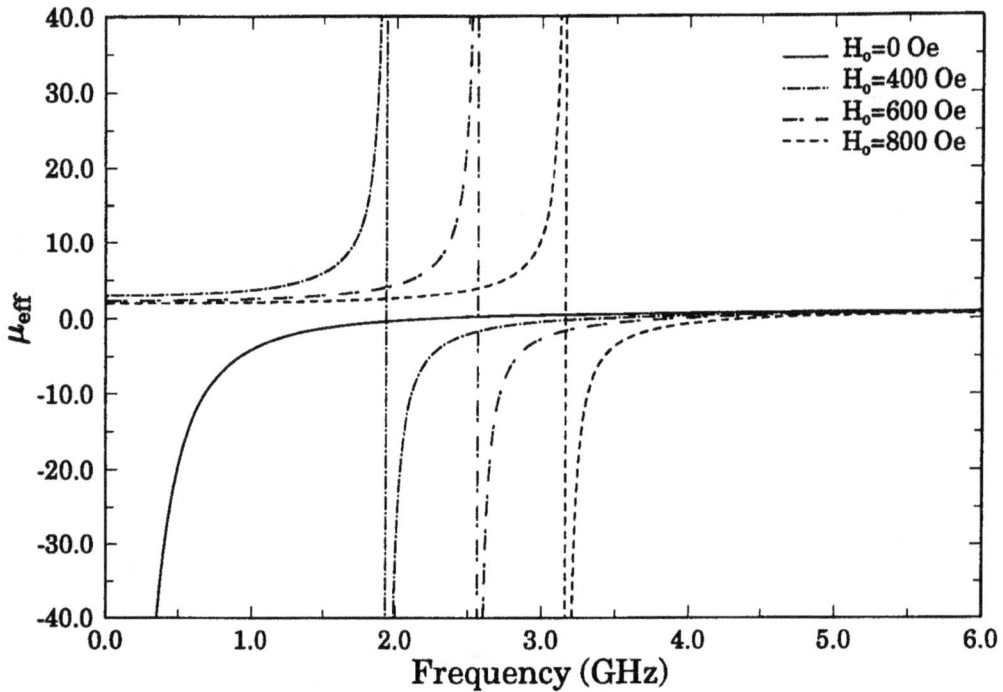

Fig. 5.20: Effective relative permeabily versus frequency as the external magnetic field varies from 0 Oe to 800 Oe. The remaining ferrite parameters are the following: $4\pi M_s = 800$ G, $\Delta H = 0$ Oe, $\epsilon_r = 13.0$.

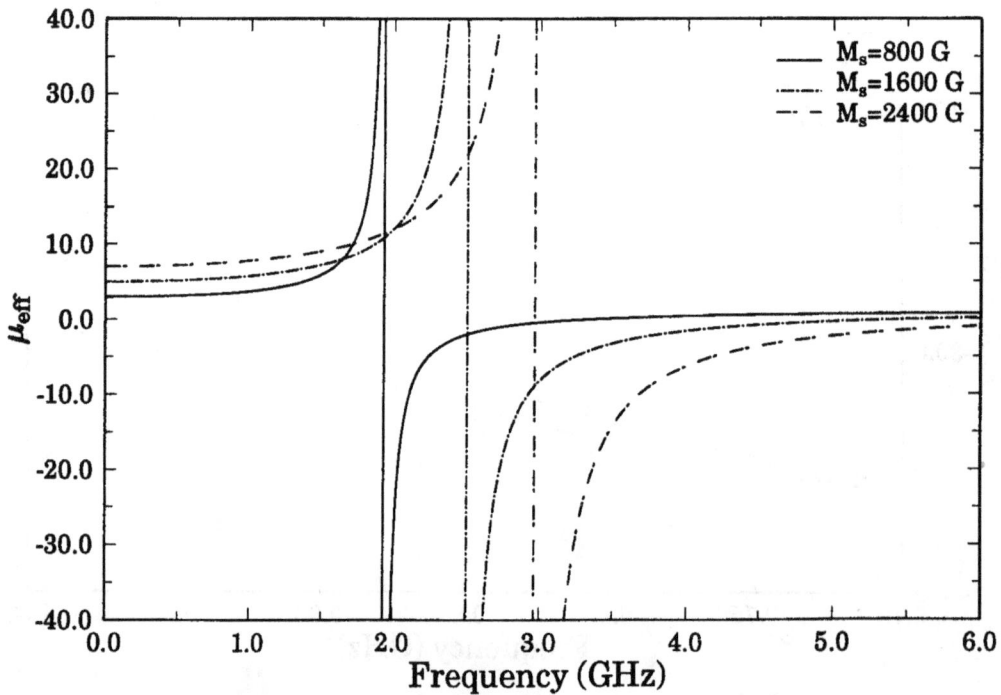

Fig. 5.21: Effective relative permeabily versus frequency as the saturization magnetization varies from 800 G to 2400 G. The remaining ferrite parameters are the following: $H_o = 400$ Oe, $\Delta H = 0$ Oe, $\epsilon_r = 13.0$.

permeability of the ferrite. The corresponding graph is illustrated in Fig. 5.21. The remaining setup parameters are kept constant. The observation is very similar to that of Fig. 5.20. The *attenuation band* constantly shifts to higher frequencies as the saturization magnetization of the ferrite increases. However, there is a small difference between the two figures: in Fig. 5.21, the three graphs actually *cross* with each other at the lower frequency range; something that does not happen in Fig. 5.20. This *crossing* effect might lead to observations where a magnetostatic resonant peak starts actually shifting to lower frequencies, instead of higher frequencies, while increasing the saturization magnetization. The exact same trend was realized when analyzing the cavity-backed aperture antenna shown in Fig. 5.16. As the saturization magnetization for the two ferrite layers increases, the first magnetostatic resonance starts shifting to a lower frequency. This observation is graphically illustrated for both polarizations in Figs. 5.22 and 5.23. Although these two figures imply a resonance shift toward the lower frequencies as the saturization magnetization increases, it will be totally untrue to claim that this is always the case. The *crossing* effect, concerning the three graphs shown in Fig. 5.20, provides a reasonable uncertainty in the direction of the frequency shift.

Another numerical experiment performed in terms of characterizing the ferromagnetic behavior of the multi-layer, cavity-backed aperture antenna, illustrated in Fig. 5.16, was to constantly vary the linewidth ΔH of the ferrite material while observing the effect on the first magnetostatic resonant peak. The results in this case were predictable. The linewidth, as was previously explained in this report, represents the lossy term of the ferrite. Thus, by increasing the linewidth of the ferrite, the level of the resonant peak is expected to significantly reduce since more energy is now dissipated inside the gyromagnetic material; however, the resonant frequency is expected to stay the same. This observation is illustrated for both vertical and horizontal polarization in Figs. 5.24 and 5.25, respectively.

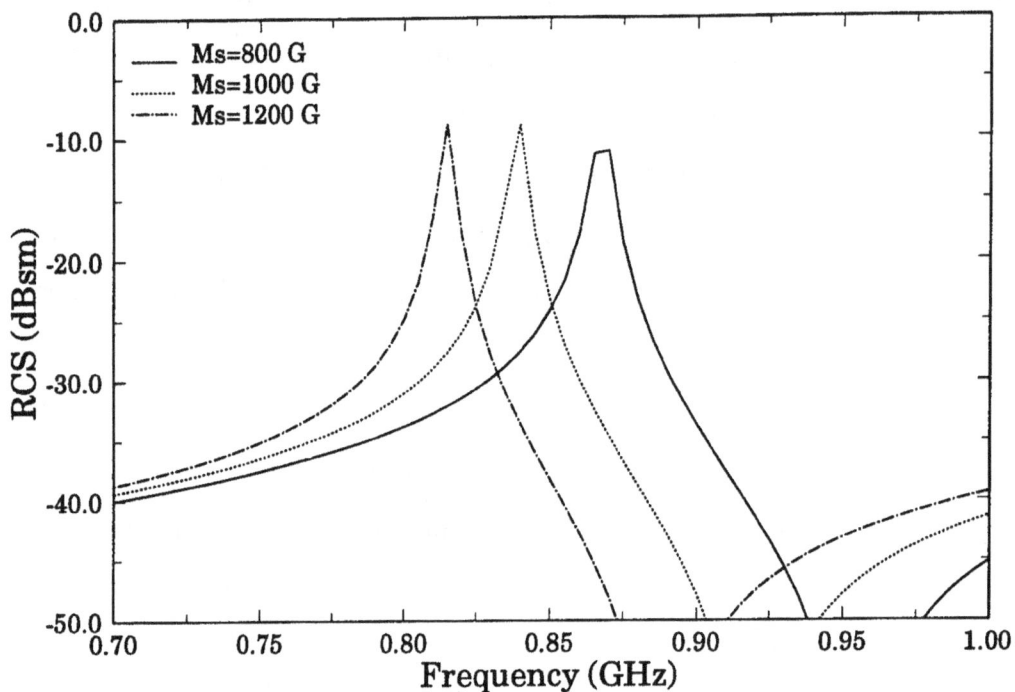

Fig. 5.22: The effect on the resonant frequency of the multi-layer antenna after changing the saturization magnetization, $4\pi M_s$; $H_o = 500$ Oe, $\Delta H = 5$ Oe, $\epsilon_r = 13.0$. $(\sigma_{\phi\phi})$.

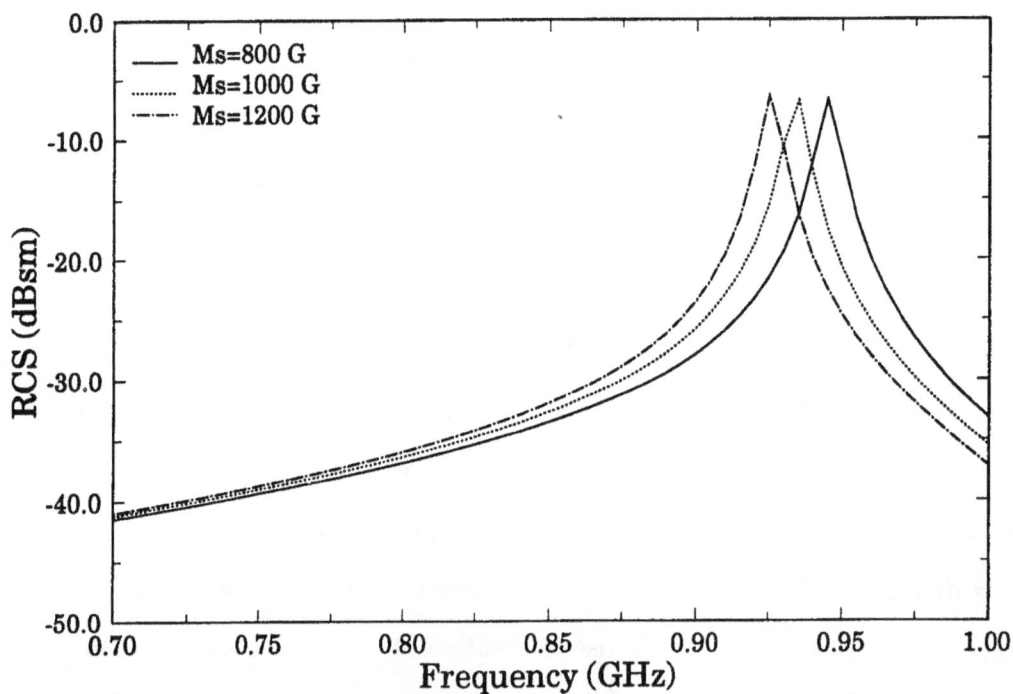

Fig. 5.23: The effect on the resonant frequency of the multi-layer antenna after changing the saturization magnetization, $4\pi M_s$; $H_o = 500$ Oe, $\Delta H = 5$ Oe, $\epsilon_r = 13.0$. $(\sigma_{\theta\theta})$.

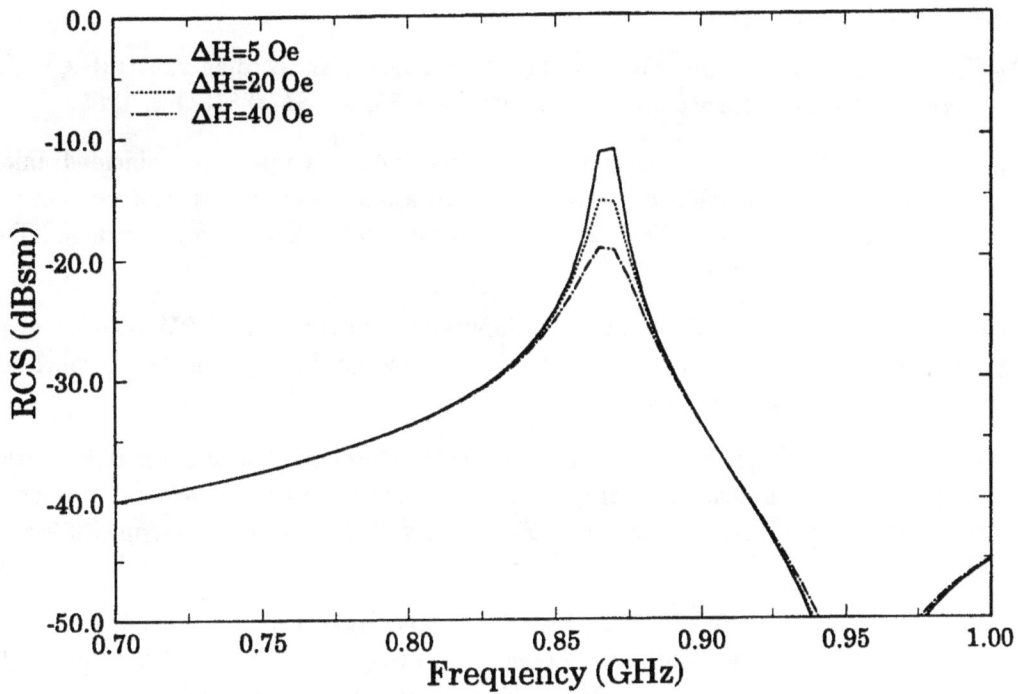

Fig. 5.24: The effect on the resonant frequency of the multi-layer antenna after increasing the linewidth (ΔH) of the ferrite material; $(\sigma_{\phi\phi})$.

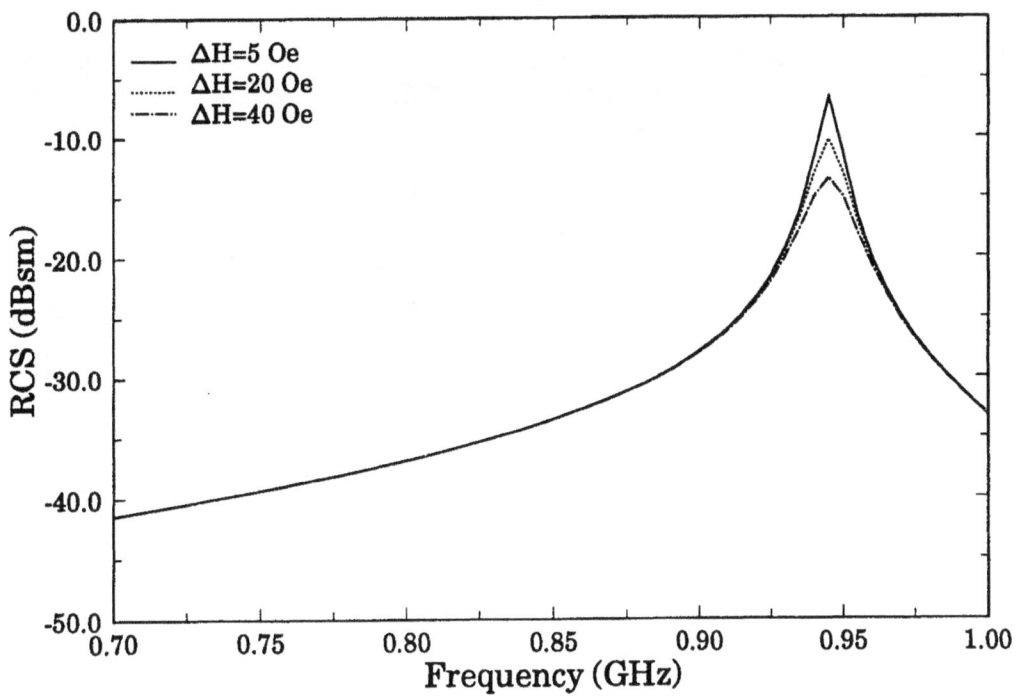

Fig. 5.25: The effect on the resonant frequency of the multi-layer antenna after increasing the linewidth (ΔH) of the ferrite material; $(\sigma_{\theta\theta})$.

REFERENCES

[1] N. G. Alexopoulos, "Integrated-circuit structures on anisotropic substrates," *IEEE Trans. Microwave Theory and Tech.*, vol. MTT-33, pp. 847–881, Oct. 1985.

[2] Y. Chen and B. Beker, "Dispersion characteristics of open and shielded microstrip lines under a combined principal axes rotation of electrically and magnetically anisotropic substrates," *IEEE Trans. Microwave Theory and Tech.*, vol. MTT-41, pp. 673–679, Apr. 1993.

[3] G. Mazé-Merceur, S. Tedjini, and J.-L. Bonnefoy, "Analysis of a CPW on electric and magnetic biaxial substrate," *IEEE Trans. Microwave Theory and Tech.*, vol. MTT-41, pp. 457–461, March 1993.

[4] A. A. Mostafa, C. M. Krowne, and K. A. Zaki, "Numerical spectral matrix method for propagation in general layered media: Application to isotropic and anisotropic substrates," *IEEE Trans. Microwave Theory and Tech.*, vol. MTT-35, pp. 1399–1407, Dec. 1987.

[5] R. R. Mansour and R. H. Macphie, "A unified hybrid-mode analysis for planar transmision lines with multilayer isotropic/anisotropic substrates," *IEEE Trans. Microwave Theory and Tech.*, vol. MTT-35, pp. 1382–1391, Dec. 1987.

[6] J. B. Davies and D. Mirshekar-Syahkal, "Spectral domain solution of arbitrary coplanar transmission line with multilayer substrate," *IEEE Trans. Microwave Theory and Tech.*, vol. MTT-25, pp. 143–146, Feb. 1977.

[7] M. Riaziat, R. Majidi-Ahy, and I. J. Feng, "Propagation modes and dispersion characteristics of coplanar waveguides," *IEEE Trans. Microwave Theory and Tech.*, vol. MTT-38, pp. 245–251, March 1990.

[8] Y. Qian, E. Yamashita, and K. Atsuki, "Modal dispersion control and distortion suppression of picosecond pulses in suspended coplanar waveguides," *IEEE Trans. Microwave Theory and Tech.*, vol. MTT-40, pp. 1903–1909, Oct. 1992.

[9] C. N. Chang, W. C. Chang, and C. H. Chen, "Full wave analysis of multilayer coplanar lines," *IEEE Trans. Microwave Theory and Tech.*, vol. MTT-39, pp. 747–750, Apr. 1991.

[10] M. R. Lyons, J. P. K. Gilb, and C. A. Balanis, "Enhanced dominant mode operation of a shielded multilayer coplanar waveguide via substrate compensation," *IEEE Trans. Microwave Theory and Tech.*, vol. MTT-41, pp. 1564–1567, Sep. 1993.

[11] T. Kitazawa and R. Mittra, "Quasi-static characteristics of asymmetrical and coupled coplanar-type transmission lines," *IEEE Trans. Microwave Theory and Tech.*, vol. MTT-33, pp. 771–778, Sep. 1985.

[12] J.-F. Lee, D. Sun, and Z. J. Cendes, "Full-wave analysis of dielectric waveguides using tangential vector finite elements," *IEEE Trans. Microwave Theory and Tech.*, vol. MTT-39, pp. 1262–1271, Aug. 1991.

[13] Y. Lu and A. Fernandez, "An efficient finite element solution of inhomogeneous anisotropic and lossy dielectric," *IEEE Trans. Microwave Theory and Tech.*, pp. 1215–1223, June/July 1993.

[14] K. Hayata, M. K., and M. Koshiba, "Finite element formulation for lossy waveguides," *IEEE Trans. Microwave Theory and Tech.*, vol. MTT-36, pp. 268–276, Feb. 1988.

[15] G. H. Golub and C. F. Van Loan, *Matrix Computations*. Baltimore: Johns Hopkins Press, 1989.

[16] I. P. Polichronakis and S. S. Kouris, "Computation of the dispersion characteristics for a shielded suspended substrate microstrip line," *IEEE Trans. Microwave Theory and Tech.*, vol. MTT-40, pp. 581–584, March 1992.

[17] Y. Rahmat-Samii, T. Itoh, and R. Mittra, "A spectral domain analysis for solving microstrip discontinuity problems," *IEEE Trans. Microwave Theory and Tech.*, vol. MTT-22, pp. 372–378, Apr. 1974.

[18] M. Helard, J. Citerne, O. Picon, and V. F. Hanna, "Theoretical and experimental investigation of finline discontinuities," *IEEE Trans. Microwave Theory and Tech.*, vol. MTT-33, pp. 994–1003, Oct. 1985.

[19] N. K. Uzunoglu, C. N. Capsalis, and C. P. Chronopoulos, "Frequency-dependent analysis of a shielded microstrip step discontinuity using an efficient mode-matching technique," *IEEE Trans. Microwave Theory and Tech.*, vol. MTT-36, pp. 976–984, June 1988.

[20] A. S. Omar and K. Schünemann, "The effect of complex modes at finline discontinuities," *IEEE Trans. Microwave Theory and Tech.*, vol. MTT-34, pp. 1508–1514, Dec. 1986.

[21] Q. Xu, K. J. Webb, and R. Mittra, "Study of modal solution procedures for microstrip step discontinuities," *IEEE Trans. Microwave Theory and Tech.*, vol. MTT-37, pp. 381–387, Feb. 1989.

[22] R. H. Jansen, "Hybrid mode analysis of end effects of planar microwave and millimetrewave transmission lines," *IEE Proceedings*, vol. 128, pt. H, pp. 77–86, Apr. 1981.

[23] J. B. Knorr, "Equivalent reactance of a shorting septum in a fin-line: Theory and experiment," *IEEE Trans. Microwave Theory and Tech.*, vol. MTT-29, pp. 1196–1202, Nov. 1981.

[24] R. H. Jansen and N. H. L. Koster, "New aspects concerning the definition of microstrip characteristic impedance as a function of frequency," in *IEEE Int. Microwave Sym.*, (Dallas, TX), pp. 305–307, 1982.

[25] J. B. Knorr and J. C. Deal, "Scattering coefficients of an inductive strip in a finline: Theory and experiment," *IEEE Trans. Microwave Theory and Tech.*, vol. MTT-33, pp. 1011–1017, Oct. 1985.

[26] N. H. L. Koster and R. H. Jansen, "The microstrip step discontinuity: A revised description," *IEEE Trans. Microwave Theory and Tech.*, vol. MTT-34, pp. 213–223, Feb. 1986.

[27] G. Schiavon, P. Tognalatti, and T. Sorrentino, "Full-wave analysis of coupled finline discontinuities," *IEEE Trans. Microwave Theory and Tech.*, vol. MTT-36, pp. 1889–1894, Dec. 1988.

[28] T. Uwano, "Accurate characterization of microstrip resonator open end with new current expression in spectral-domain approach," *IEEE Trans. Microwave Theory and Tech.*, vol. MTT-37, pp. 630–633, March 1989.

[29] D. M. Pozar, "Input impedance and mutual coupling of rectangular microstrip antennas," *IEEE Trans. Antennas and Propgat.*, vol. AP-30, pp. 1191–1196, Nov. 1982.

[30] R. W. Jackson and D. M. Pozar, "Full-wave analysis of microstrip open-end and gap discontinuities," *IEEE Trans. Microwave Theory and Tech.*, vol. MTT-33, pp. 1036–1042, Oct. 1985.

[31] Q. Zhang and T. Itoh, "Spectral-domain analysis of scattering from E-plane circuit elements," *IEEE Trans. Microwave Theory and Tech.*, vol. MTT-35, pp. 138–150, Feb. 1987.

[32] J. P. Gilb, *Transient Signal Analysis of Multilayer Multiconductor Microstrip Transmission Lines.* Master's thesis, Arizona State University, 1989.

[33] T. Itoh, "Spectral domain immitance approach for dispersion characteristics of generalized printed transmission lines," *IEEE Trans. Microwave Theory and Tech.*, vol. MTT-28, pp. 733–736, July 1980.

[34] J. P. Gilb and C. A. Balanis, "Pulse distortion on multilayer coupled microstrip lines," *IEEE Trans. Microwave Theory and Tech.*, vol. MTT-37, pp. 1620–1628, Oct. 1989.

[35] M. R. Lyons and C. A. Balanis, "Transient coupling reduction and design considerations in edge-coupled coplanar waveguide couplers," *IEEE Trans. Microwave Theory and Tech.*, vol. MTT-44, pp. 778–783, May 1996.

[36] J. P. K. Gilb and C. A. Balanis, "Asymmetric, multi-conductor, low-coupling structures for high-speed, high-density digital interconnects," *IEEE Trans. Microwave Theory and Tech.*, vol. MTT-39, pp. 2100–2106, Dec. 1991.

[37] D. Kajfez, "Raise coupler directivity with lumped compensation," *Microwaves*, vol. 19, pp. 64–70, March 1976.

[38] D. D. Paolino, "MIC overlay coupler design using spectral domain," *IEEE Trans. Microwave Theory and Tech.*, vol. MTT-26, pp. 646–649, Sep. 1978.

[39] L. Su, T. Itoh, and J. Rivera, "Design of an overlay directional coupler by a full-wave analysis," *IEEE Trans. Microwave Theory and Tech.*, vol. MTT-31, pp. 1017–1022, Dec. 1983.

[40] L. Carin and K. J. Webb, "Isolation effects in single- and dual-plane VLSI interconnects," *IEEE Trans. Microwave Theory and Tech.*, vol. MTT-38, pp. 396–404, Apr. 1990.

[41] S. Seki and H. Hasegawa, "Analysis of crosstalk in very high-speed LSI/VLSI's using a coupled multi-conductor MIS microstrip line model," *IEEE Trans. Microwave Theory and Tech.*, vol. MTT-32, pp. 1715–1720, Dec. 1984.

[42] M. Kirschning and R. H. Jansen, "Accurate model for effective dielectric constant of microstrip with validity up to millimetre-wave frequencies," *Electronic Letters*, vol. 18, pp. 272–273, March 1982.

[43] M. Kirschning and R. H. Jansen, "Accurate wide-range design equations for the frequency-dependent characteristics of parallel coupled microstrip lines," *IEEE Trans. Microwave Theory and Tech.*, vol. MTT-32, pp. 83–90, Jan. 1984.

[44] Y. Y. Wang, G. L. Wang, and Y. H. Shu, "Analysis and synthesis equations for edge-coupled suspended substrate microstrip line," in *IEEE Int. Microwave Sym.*, (Long Beach, CA), pp. 1123–1126, 1989.

[45] T. Uwano and T. Itoh, "Spectral domain approach," in *Numerical Techniques for Microwave and Millimeter-Wave Passive Structures*, (T. Itoh, ed.), ch. 5, pp. 334–380, New York: John Wiley and Sons, 1989.

[46] J. P. K. Gilb and C. A. Balanis, "MIS slow-wave structures over a wide range of parameters," in *IEEE Int. Microwave Sym.*, (Albuquerque, NM), pp. 877–880, 1992.

[47] K. S. Yee, "Numerical solution of initial boundary value problems involving Maxwell's equations in isotropic media," *IEEE Trans. Antennas Propagat.*, pp. 302–307, May 1966.

[48] R. W. Jackson, "Full-wave, finite element analysis of irregular microstrip discontinuities," *IEEE Trans. Microwave Theory and Tech.*, vol. MTT-37, pp. 81–89, Jan. 1989.

[49] P. Silvester, "Finite element solution of homogeneous waveguide problems," *Alta Frequenza*, vol. 38, pp. 313–317, May 1969.

[50] X. Zhang and K. K. Mei, "Time-domain finite difference approach to the calculation of frequency-dependent characteristics of microstrip discontinuities," *IEEE Trans. Microwave Theory and Tech.*, vol. MTT-36, pp. 1775–1787, Dec. 1988.

[51] X. Zhang, J. Fang, K. K. Mei, and Y. Liu, "Calculations of the dispersive characteristics of microstrips by the time-domain finite difference method," *IEEE Trans. Microwave Theory and Tech.*, vol. MTT-36, pp. 263–267, Feb. 1988.

[52] D. M. Sheen, S. M. Ali, M. D. Abouzahra, and J. A. Kong, "Application of the three-dimensional finite-difference time-domain method to the analysis of planar microstrip circuits," *IEEE Trans. Microwave Theory and Tech.*, vol. MTT-38, pp. 849–857, July 1990.

[53] A. Christ and H. L. Hartnagel, "Three-dimensional finite-difference method for the analysis of microwave-device embedding," *IEEE Trans. Microwave Theory and Tech.*, vol. MTT-35, pp. 688–696, Aug. 1987.

[54] T. Becks and I. Wolff, "Analysis of 3-D metallization structures by a full-wave spectral domain technique," *IEEE Trans. Microwave Theory and Tech.*, vol. MTT-40, pp. 2219–2227, Dec. 1992.

[55] J. C. Nedelec, "Mixed finite elements in R^3," *Numer. Meth.*, vol. 35, pp. 315–341, 1980.

[56] A. F. Peterson, "Absorbing boundary conditions for the vector wave equation," *Microwave Opt. Tech. Lett.*, vol. 1, pp. 62–64, Apr. 1988.

[57] J. Jin, *The Finite Element Method in Electromagnetics.* New York: Wiley, 1993.

[58] K. Ise, K. Inoue, and M. Koshiba, "Three-dimensional finite-element method with edge elements for electromagnetic waveguide discontinuities," *IEEE Trans. Microwave Theory and Tech.*, vol. MTT-39, pp. 1289–1295, Aug. 1991.

[59] A. C. Polycarpou, M. R. Lyons, and C. A. Balanis, "Finite element analysis of MMIC waveguide structures with anisotropic substrates," *IEEE Trans. Microwave Theory and Tech.*, vol. MTT-44, No. 10, pp. 1650–1663, Oct. 1996.

[60] J.-S. Wang and R. Mittra, "Finite element analysis of MMIC structures and electronic packages using absorbing boundary conditions," *IEEE Trans. Microwave Theory and Tech.*, vol. MTT-42, pp. 441–449, March 1994.

[61] J.-S. Wang and R. Mittra, "A finite element cavity resonance method for waveguide and microstrip line discontinuity problems," *IEEE Trans. Microwave Theory and Tech.*, vol. MTT-42, pp. 433–440, March 1994.

[62] J.-G. Yook, D. I. Nihad, and L. P. B. Katehi, "Characterization of high frequency interconnects using finite difference time domain and finite element methods," *IEEE Trans. Microwave Theory and Tech.*, vol. MTT-42, pp. 1727–1736, Sep. 1994.

[63] E. Pettenpaul, H. Kapusta, A. Weisgerber, H. Mampe, J. Luginsland, and I. Wolff, "CAD models of lumped elements on GaAs up to 18 GHz," *IEEE Trans. Microwave Theory and Tech.*, vol. MTT-36, pp. 294–304, Feb. 1988.

[64] L. Vietzorreck and R. Pregla, "Hybrid analysis of three-dimensional MMIC elements by the method of lines," *IEEE Trans. Microwave Theory and Tech.*, vol. MTT-44, pp. 2580–2586, Dec. 1996.

[65] J. P. Berenger, "A perfectly matched layer for the absorption of electromagnetic waves," *J. Comp. Phys.*, vol. 114, no. 2, pp. 185–200, Oct. 1994.

[66] Z. S. Sacks, D. M. Kingsland, R. Lee, and J.-F. Lee, "A perfectly matched anisotropic absorber for use as an absorbing boundary condition," *IEEE Trans. Antennas Propagat.*, vol. 43, no. 12, pp. 1460–1463, Dec. 1995.

[67] A. C. Polycarpou, M. R. Lyons, and C. A. Balanis, "A two-dimensional finite element formulation of the perfectly matched layer," *IEEE Microwave and Guided Wave Lett.*, vol. vol. 6, pp. 338–340, Sep. 1996.

[68] D. M. Pozar, "Improved computational efficiency for the method of moments solution of printed dipoles and patch," *Electromagnetics*, vol. 3, pp. 299–309, 1983.

[69] J. R. Mosig and F. E. Gardiol, "A dynamical radiation model for microstrip structures," *Advances in Electronics and Electron Physics*, vol. 59, pp. 139–161, 1982.

[70] R. W. Jackson, "Considerations in the use of coplanar waveguide for millimeter-wave integrated circuits," *IEEE Trans. Microwave Theory and Tech.*, vol. MTT-34, pp. 1450–1456, Dec. 1986.

[71] T. Itoh and R. Mittra, "Spectral domain approach for calculating the dispersion characteristics of microstrip lines," *IEEE Trans. Microwave Theory and Tech.*, vol. MTT-21, pp. 496–499, July 1973.

[72] J. P. Gilb, *Transient Signal Analysis of Multilayer Multiconductor Microstrip Transmission Lines*. Master's thesis, Arizona State University, 1989.

[73] N. Fache and D. D. Zutter, "Rigorous full-wave space-domain solution for dispersive microstrip lines," *IEEE Trans. Microwave Theory and Tech.*, vol. MTT-36, pp. 731–737, Apr. 1988.

[74] J. S. Bagby, C.-H. Lee, D. P. Nyquist, and Y. Yuan, "Identification of propagation regimes on integrated microstrip transmission lines," *IEEE Trans. Microwave Theory and Tech.*, vol. MTT-41, pp. 1887–1894, Nov. 1993.

[75] S.-O. Park and C. A. Balanis, "Dispersion characteristics of open microstrip lines using closed-form asymptotic extraction," *IEEE Trans. Microwave Theory and Tech.*, vol. MTT-45, March 1997.

[76] P. B. Katéhi and N. G. Alexópoulos, "Real axis integration of Sommerfeld integrals with application to printed circuit antennas," *J. Math. Phys.*, vol. 24, pp. 527–533, 1983.

[77] J. R. Mosig and F. E. Gardiol, "Analytical and numerical techniques in the Green's function treatment of microstrip antennas and scatterers," *Inst. Elec. Eng. Proc.*, vol. 130, pp. 175–183, March 1983.

[78] J. R. Mosig and T. K. Sarkar, "Comparison of quasi-static and exact electromagnetic fields from a horizontal electric dipole above a lossy dielectric backed by an imperfect ground," *IEEE Trans. Microwave Theory and Tech.*, vol. MTT-34, pp. 379–387, 1986.

[79] S. Barkeshli, P. H. Pathak, and M. Martin, "An asymptotic closed-form microstrip surface Green's function for the efficient moment method analysis of mutual coupling in microstrip antennas," *IEEE Trans. Antennas and Propagat.*, vol. AP-38, pp. 1374–1383, Sep. 1990.

[80] Y. L. Chow, J. J. Yang, D. H. Fang, and G. E. Howard, "Closed-form spatial Green's function for the thick substrate," *IEEE Trans. Microwave Theory and Tech.*, vol. MTT-39, pp. 588–592, March 1991.

[81] M. I. Aksun and R. Mittra, "Derivation of closed-form Green's functions for a general microstrip geometry," *IEEE Trans. Microwave Theory and Tech.*, vol. MTT-40, pp. 2055–2062, Nov. 1992.

[82] M. Kobayashi and F. Ando, "Dispersion characteristics of open microstrip lines," *IEEE Trans. Microwave Theory and Tech.*, vol. MTT-35, pp. 784–788, Feb. 1987.

[83] M. Kobayashi and T. Iijima, "Frequency-dependent characteristics of current distributions on microstrip lines," *IEEE Trans. Microwave Theory and Tech.*, vol. MTT-37, pp. 799–801, Apr. 1989.

[84] C. Shih, R.-B. Wu, and C. H. Chen, "A full-wave analysis of microstrip lines by variational conformal mapping technique," *IEEE Trans. Microwave Theory and Tech.*, vol. MTT-36, pp. 576–581, March 1988.

[85] C. A. Balanis, *Advanced Engineering Electromagnetics*. New York: John Wiley and Sons, 1989.

[86] S.-O. Park and C. A. Balanis, "Closed-form asymptotic extraction method for coupled microstrip lines," *IEEE Trans. on Microwave and Guided Wave Letter*, vol. MWG-7, March 1997.

[87] G. Kowalski and R. Pregla, "Dispersion characteristics of single and coupled microstrips," *Arch. Elek. Übertrang.*, vol. 26, pp. 276–280, 1972.

[88] S.-O. Park and C. A. Balanis, "Analytical transform technique to evaluate the asymptotic part of impedance matrix of Sommerfeld-type integrals," *IEEE Trans. Antennas and Propagat.*, vol. AP-45, May 1997.

[89] M. Marin, S. Barkeshli, and P. H. Pathak, "Efficient analysis of planar microstrip geometries using a closed-form asymptotic representation of the grounded dielectric slab Green's function," *IEEE Trans. Microwave Theory and Tech.*, vol. MTT-37, pp. 669–679, Apr. 1989.

[90] T. R. Arabi, A. T. Murphy, T. K. Sarkar, R. F. Harrington, and A. R. Djordjević, "Analysis of arbitrarily oriented microstrip lines utilizing a quasi-dynamic approach," *IEEE Trans. Microwave Theory and Tech.*, vol. MTT-39, pp. 75–82, Jan. 1991.

[91] S.-O. Park and C. A. Balanis, "Analytical evaluation of the asymptotic impedance matrix of asymmetric gap discontinuities," 1997. Preparation for Submission in *IEEE Trans. Microwave Theory and Tech.*

[92] S.-O. Park, *Analytical Techniques for the Evaluation of Asymptotic Matrix Elements in Electromagnetic Problems*. PhD thesis, Arizona State University, May 1997.

[93] G. W. G. Pan, J. Tan, and J. D. Murphy, "Full-wave analysis of microstrip floating-line discontinuities," *IEEE Trans. Electromag. Compatib.*, vol. 36, pp. 49–59, Feb. 1994.

[94] *The Microwave Engineer's Handbook and Buyers' Guide.* Horizon House: New York, 1969, p72.

[95] N. G. Alexópoulos and S.-C. Wu, "Frequency-independent equivalent circuit model for microstrip open-end and gap discontinuities," *IEEE Trans. Microwave Theory and Tech.*, vol. MTT-42, pp. 1268–1272, July 1994.

[96] M. C. Bailey and M. D. Deshpande, "Integral equation formulation of microstrip antennas," *IEEE Trans. Antennas and Propagat.*, vol. AP-30, pp. 651–656, July 1982.

[97] E. H. Newman and D. Forrai, "Scattering from a microstrip patch," *IEEE Trans. Antennas and Propagat.*, vol. AP-35, pp. 245–251, March 1987.

[98] S.-O. Park, C. A. Balanis, and C. R. Birtcher, "Analytical evaluation of the asymptotic impedance matrix of a grounded dielectric slab with roof-top function," 1997. Submitted for review in *IEEE Trans. Antenna Propagat.*

[99] T. S. Horng, W. E. McKinzie, and N. G. Alexópoulos, "Full-wave spectral-domain analysis of compensation of microstrip discontinuities using triangular subdomain functions," *IEEE Trans. Microwave Theory and Tech.*, vol. MTT-40, pp. 2137–2147, Dec. 1992.

[100] J. T. Aberle. Arizona State University, Private Communication.

[101] F. Zavosh and J. T. Aberle, "Infinite phased arrays of cavity-backed patches," *IEEE Trans. Antennas Propagat.*, vol. 42, pp. 390–398, March 1994.

[102] J. T. Aberle and F. Zavosh, "Analysis of probe-fed circular microstrip patches backed by circular cavities," *Electromagnetics*, vol. 14, pp. 239–258, Apr. 1994.

[103] S. M. Rao, D. R. Wilton, and A. W. Glisson, "Electromagnetic scattering by surfaces of arbitrary shape," *IEEE Trans. on Antennas and Propagation*, vol. AP-30, pp. 409–418, May 1982.

[104] *Structural Dynamics Research Corporation, SDRC I-DEAS.*

[105] F. Zavosh, *Analysis of Circular Microstrip Patch Antennas Backed by Circular Cavities.* Master's thesis, Arizona State University, 1993.

[106] F. Zavosh, *Novel Printed Antenna Configurations for Enhanced Performance.* PhD thesis, Arizona State University, 1995.

[107] J. M. Jin and J. L. Volakis, "A hybrid finite element method for scattering and radiation by microstrip patch antennas and arrays residing in a cavity," *IEEE Trans. Antennas Propagat.*, vol. AP-39, pp. 1598–1604, Nov. 1991.

[108] H. L. Glass, "Ferrite films for microwave and millimeter wave devices," *IEEE Proceedings*, vol. 76, pp. 151–158, Feb. 1988.

[109] D. M. Pozar, "Radiation and scattering characteristics of microstrip antennas on normally biased ferrite substrates," *IEEE Trans. Antennas Propagat.*, vol. AP-40, pp. 1084–1092, Sep. 1992.

[110] D. M. Pozar, "Correction to radiation and scattering characteristics of microstrip antennas on normally biased ferrite substrates," *IEEE Trans. Antennas Propagat.*, vol. AP-42, pp. 122–123, Jan. 1994.

[111] D. Kokotoff, *Full-Wave Analysis of a Ferrite-Tuned Cavity-Backed Slot Antenna.* PhD thesis, Arizona State University, 1995.

[112] D. M. Pozar and V. Sanchez, "Magnetic tuning of a microstrip antenna on a ferrite substrate," *Electron. Lett.*, vol. vol. 24, pp. 729–731, June 1988.

[113] D. M. Pozar, "Radar cross-section of a microstrip antenna on normally biased ferrite substrate," *Electron. Lett.*, vol. vol. 25, no. 16, pp. 1079–1080, Aug. 1989.

[114] H.-Y. Yang, "Characteristics of switchable ferrite microstrip antennas," *IEEE Trans. Antennas and Propagat.*, vol. AP-44, no. 8, pp. 1127–1132, Aug. 1996.

[115] D. M. Pozar, *Microwave Engineering.* Massachusetts: Addison-Wesley Publishing Company, Inc., 1990.